MILLAN
WORK OUT
SERIES

Work Out

Applied Mathematics

'A' Level

The titles
in this
series

MACMILLAN
WORK OUT
SERIES

Work Out

Applied Mathematics

'A' Level

R. Haines

and

B. Haines

MACMILLAN

First published 1986
Reprinted (with corrections) 1987

Published by
MACMILLAN EDUCATION LTD
Houndmills, Basingstoke, Hampshire RG21 2XS
and London
Companies and representatives
throughout the world

Typeset and illustrated by TecSet Ltd,
Wallington, Surrey
Printed in Great Britain by The Bath Press, Avon

British Library Cataloguing in Publication Data
Haines, R.
Work out applied mathematics 'A' level.—
(Macmillan Work out series)
1. Mathematics—Examinations, questions, etc.
I. Title   II. Haines, B.
510'.76   QA43
ISBN 0-333-39767-3

Contents

To Kevin, Sarah and Nicholas

Acknowledgements

Over many years the questions set by the various Examination Boards have stimulated and enhanced the teaching of mathematics throughout education. Everyone involved in mathematics, both the teachers and the taught, owes a debt to the Boards for the ever-present challenge that new examination questions bring to mathematics education.

Once again our thanks go to everyone who has helped with the preparation of this book, especially to Nicholas who, with nimble fingers, processed our indifferent typing into an acceptable form.

We shall be greatly indebted to anyone notifying us of any errors.

The author and publishers wish to thank the following who have kindly given permission for the use of copyright material:

The Associated Examining Board, the Southern Universities' Joint Board, the University of London School Examinations Board and the University of Oxford Delegacy of Local Examinations for questions from past examination papers.

Every effort has been made to trace all the copyright holders but if any have been inadvertently overlooked the publishers will be pleased to make the necessary arrangement at the first opportunity.

Examination Boards for Advanced level

Syllabuses and past examination papers can be obtained from:

The Associated Examining Board (AEB)
Stag Hill House
Guildford
Surrey GU2 5XJ

University of Cambridge Local Examinations Syndicate (UCLES)
Syndicate Buildings
Hills Road
Cambridge CB1 2EU

Joint Matriculation Board (JMB)
78 Park Road
Altrincham
Cheshire WA14 5QQ

University of London School Examinations Board (L)
University of London Publications Office
52 Gordon Square London WC1E 6EE

University of Oxford Delegacy of Local Examinations (OLE)
Ewert Place
Summertown
Oxford OX2 7BZ

Oxford and Cambridge Schools Examination Board (O & C)
10 Trumpington Street
Cambridge CB2 1QB

Scottish Examination Board (SEB)
Robert Gibson & Sons (Glasgow) Ltd
17 Fitzroy Place Glasgow G3 7SF

Southern Universities' Joint Board (SUJB)
Cotham Road
Bristol BS6 6DD

Welsh Joint Education Committee (WJEC)
245 Western Avenue
Cardiff CF5 2YX

Northern Ireland Schools Examination Council (NISEC)
Examinations Office
Beechill House
Beechill Road
Belfast BT8 4RS

Introduction

The Work Out series are not 'just textbooks'. They are based on the revision needs of 'A' level students and have been designed to help students obtain the best possible grades in their examinations.

This applied mathematics book is based on the mechanics and probability content of 'single subject' mathematics syllabuses. Coupled with *Work Out Pure Mathematics 'A' Level* by the same authors, most of the topics likely to be encountered in 'A' level mathematics examinations have been covered.

Additionally, students entering higher education to read mathematics and associated topics often find they are unfamiliar with topics which their new lecturers assume they have already covered. Typically, students who have studied mathematics and/or statistics for 'A' level find themselves at a loss when their lecturer embarks on dynamics. This book provides a bridge which will allow students to rapidly familiarise themselves, through examples, with standard topics of applied mathematics.

Each chapter in the book starts with a brief list of formulae and is followed by many 'A' level type questions, each with a complete solution. At the end of each chapter more 'A' level questions are set as exercises with the important steps in the working of each one being given to facilitate easy reference.

How to Use the Book

(a) By repeated use and practice, endeavour to become familiar with the frequently used formulae listed in the Fact Sheets.
(b) Practice in answering examination questions is important. Open the book at a definite topic, choose a question and cover over the solution until you have tried to do it by yourself. If you get really stuck your mind will be receptive when you uncover the solution.
(c) The methods used in the book are not necessarily the shortest. Always be on the lookout for shorter and neater solutions. 'There is always a shorter way' is an excellent maxim to adopt throughout mathematics.

Revision

(a) Your school or college should be able to supply you with a syllabus and typical examination papers. If not, you should write to the secretary of the examination board. A list of addresses is given on pages vii and viii.
(b) Use the book to revise topics before trying past papers.
(c) Familiarise yourself with the contents of your formula booklet (if one is provided) well before the examination. There are always some candidates who fail to answer questions because they are not aware that a vital formula which escapes them at that instant is to be found in the booklet so thoughtfully provided by the board!

(d) Develop a good examination technique. Try papers 'to time' under examination conditions (quiet, no cups of coffee!). Don't be too depressed by your first efforts; practice can make perfect.

(e) Get into the habit of writing solutions tidily first time. It makes it so much easier on 'the day' if you don't have to think about legibility.

The Examination

(a) Spend a few minutes reading the paper and select questions in order of priority.

(b) Do the 'easy' questions at the start of the examination — this boosts your confidence.

(c) If you get stuck on a question and cannot see an alternative approach, cut your losses and go on to another question. Examinations have been failed by good candidates who spent too long stuck on one or two questions.

(d) Never cross out. You may be crossing out marks.

(e) Never walk out of an examination. Reread the questions right through, even if you cannot do the first part of a question, there may be parts you can try.

(f) Good luck.

1 Vector Dynamics

Differentiation and integration of two- and three-dimensional vectors with respect to a scalar variable. Displacement, velocity and acceleration as vectors. The equations of motion of a particle in vector form.

1.1 Fact Sheet

You are advised to read the chapter on vectors in *Work Out Pure Mathematics 'A' Level*. In vector work, always remember that one vector equation may be expressed as two or three scalar equations (in two or three dimensions).

(a) Position Vectors

If O is an origin then $\overline{OP}$ is the position vector of a point P relative to O; a common notation is $\overline{OP} = \mathbf{r}$.

(b) Displacement Vectors

If $\overline{OP} = \mathbf{r}_1$, and $\overline{OQ} = \mathbf{r}_2$, then $\overline{PQ}$ is the displacement vector from P to Q, and $\overline{PQ} = \mathbf{r}_2 - \mathbf{r}_1$.

(c) Cartesian Components

A point P with coordinates (x, y, z) has position vector

$$\overline{OP} = \mathbf{r} = x\mathbf{i} + y\mathbf{j} + z\mathbf{k} = \begin{pmatrix} x \\ y \\ z \end{pmatrix}.$$

The distance OP, r, is the magnitude of the vector $\mathbf{r}$;

$$r = |\mathbf{r}| = \sqrt{(x^2 + y^2 + z^2)}.$$

(d) Scalar Product

The scalar product of the vectors

$$\mathbf{r}_1 = x_1\mathbf{i} + y_1\mathbf{j} + z_1\mathbf{k} \qquad \text{and} \qquad \mathbf{r}_2 = x_2\mathbf{i} + y_2\mathbf{j} + z_2\mathbf{k} \qquad \text{is defined as}$$

$$\mathbf{r}_1 \cdot \mathbf{r}_2 = r_1 r_2 \cos\theta,$$

where θ is the angle between $\mathbf{r}_1$ and $\mathbf{r}_2$.

$$\mathbf{r}_1 \cdot \mathbf{r}_2 = x_1 x_2 + y_1 y_2 + z_1 z_2; \qquad \mathbf{r} \cdot \mathbf{r} = r^2 = x^2 + y^2 + z^2.$$

(e) Velocity and Acceleration

Position vector $\overline{OP} = \mathbf{r}(t) = x(t)\mathbf{i} + y(t)\mathbf{j} + z(t)\mathbf{k}$.

Velocity vector $= \mathbf{v} = \dfrac{d\mathbf{r}}{dt} = \dfrac{dx}{dt}\mathbf{i} + \dfrac{dy}{dt}\mathbf{j} + \dfrac{dz}{dt}\mathbf{k}$,

$\quad$ or $\quad \mathbf{v} = \dot{\mathbf{r}} = \dot{x}\mathbf{i} + \dot{y}\mathbf{j} + \dot{z}\mathbf{k}$.

Speed $= |\mathbf{v}| = \sqrt{(\dot{x}^2 + \dot{y}^2 + \dot{z}^2)}$.

Acceleration vector $\mathbf{a} = \dfrac{d\mathbf{v}}{dt} = \dfrac{d^2\mathbf{r}}{dt^2} = \dfrac{d^2x}{dt^2}\mathbf{i} + \dfrac{d^2y}{dt^2}\mathbf{j} + \dfrac{d^2z}{dt^2}\mathbf{k}$

$\quad$ or $\mathbf{a} = \dot{\mathbf{v}} = \ddot{\mathbf{r}} = \ddot{x}\mathbf{i} + \ddot{y}\mathbf{j} + \ddot{z}\mathbf{k}$.

(f) Equations of Motion

$\mathbf{F} = m\mathbf{a}$, $\qquad$ or $\qquad F_x\mathbf{i} + F_y\mathbf{j} + F_z\mathbf{k} = m\,(\ddot{x}\mathbf{i} + \ddot{y}\mathbf{j} + \ddot{z}\mathbf{k})$.

If $\mathbf{a}$ is constant then
$\mathbf{v} = \mathbf{u} + \mathbf{a}t \qquad$ where $\mathbf{u}$ is the initial velocity,
$\mathbf{r} = \mathbf{r}_0 + \mathbf{u}t + \frac{1}{2}\mathbf{a}t^2 \qquad$ where $\mathbf{r}_0$ is the initial position vector.

When $\mathbf{a} = 0 \qquad$ then $\qquad \mathbf{v} = \mathbf{u}$ (constant) $\qquad$ and $\mathbf{r} = \mathbf{r}_0 + \mathbf{u}t$.

Note 1 $\quad$ When integrating a vector equation remember that the constant of integration is also a vector.
Note 2 $\quad$ In two dimensions the direction of motion is $\arctan\left(\dfrac{\dot{y}}{\dot{x}}\right)$ with the positive x-axis.

1.2 Worked Examples

1.1 $\quad$ Simplify $4\overline{BC} + \overline{AB} + 3\overline{DB} + 3\overline{CD}$, where $ABCD$ is a plane quadrilateral.

- $3\overline{DB} + 3\overline{CD} = 3\,(\overline{CD} + \overline{DB}) = 3\overline{CB}$
$$= -3\overline{BC}.$$
Hence:
$$4\overline{BC} + \overline{AB} + 3\overline{DB} + 3\overline{CD} = 4\overline{BC} + \overline{AB} - 3\overline{BC}$$
$$= \overline{AB} + \overline{BC}$$
$$= \overline{AC}.$$

1.2 $\quad$ A particle P has position vector $\mathbf{r}$ at time t where
$$\mathbf{r} = 2\,(1 + \cos t)\mathbf{i} + 2\,(t - \sin t)\mathbf{j} + 3\mathbf{k}.$$

Show that the speed of the particle at time t is $4\left|\sin\dfrac{t}{2}\right|$ and that the acceleration of the particle has constant magnitude.

- If $\mathbf{r} = 2\,(1 + \cos t)\mathbf{i} + 2\,(t - \sin t)\mathbf{j} + 3\mathbf{k}$
then velocity $= \dot{\mathbf{r}} = -2\sin t\mathbf{i} + 2\,(1 - \cos t)\mathbf{j} + 0\mathbf{k}$
and acceleration $= \ddot{\mathbf{r}} = -2\cos t\mathbf{i} + 2\sin t\mathbf{j} + 0\mathbf{k}$.

$$(\text{Speed})^2 = |\dot{\mathbf{r}}|^2 = 4\sin^2 t + 4(1 - \cos t)^2$$
$$= 4(\sin^2 t + 1 + \cos^2 t - 2\cos t)$$
$$= 8(1 - \cos t)$$
$$= 16\sin^2 \frac{t}{2},$$

hence $\qquad$ speed $= 4\left|\sin\dfrac{t}{2}\right|$.

$$(\text{Acceleration})^2 = 4\cos^2 t + 4\sin^2 t$$
$$= 4,$$
hence $|\text{acceleration}| = 2$.

1.3 A particle P of unit mass, moving under gravity, has position vector $\mathbf{r}$ at time t, where $\dfrac{d^2\mathbf{r}}{dt^2} = \mathbf{g}$. If the particle is at the origin with velocity $\mathbf{u}$ when $t = 0$, show that

$$\frac{d\mathbf{r}}{dt} = \mathbf{g}t + \mathbf{u} \qquad \text{and} \qquad \mathbf{r} = \mathbf{u}t + \mathbf{g}\frac{t^2}{2}.$$

- $\dfrac{d^2\mathbf{r}}{dt^2} = \mathbf{g}$.

Integrating with respect to time, $\displaystyle\int \frac{d^2\mathbf{r}}{dt^2}\,dt = \int \mathbf{g}\,dt$, $\qquad$ i.e. $\qquad \dfrac{d\mathbf{r}}{dt} = \mathbf{g}t + \mathbf{A}$.

But $\dfrac{d\mathbf{r}}{dt} = \text{velocity} = \mathbf{u}$ when $t = 0$, so $\mathbf{A} = \mathbf{u}$; $\qquad \dfrac{d\mathbf{r}}{dt} = \mathbf{g}t + \mathbf{u}$.

Integrating again, $\qquad \mathbf{r} = \mathbf{g}\dfrac{t^2}{2} + \mathbf{u}t + \mathbf{B}$.

But $\mathbf{r} = 0$ when $t = 0$, so $\mathbf{B} = 0$,

$\mathbf{r} = \mathbf{u}t + \mathbf{g}\dfrac{t^2}{2}$.

1.4 (multiple choice) A particle moves so that its position vector at time t seconds is given by $\mathbf{r}$ metres, relative to a fixed origin O, where $\mathbf{r} = 2t^3\mathbf{i} - 9t\mathbf{j}$.
The speed in m s^{-1} of the particle when $t = 1$ is

A, 85; $\qquad$ B, -3; $\qquad$ C, $-3\sqrt{2}$; $\qquad$ D, $3\sqrt{13}$; $\qquad$ E, none of these.

- $\mathbf{r} = 2t^3\mathbf{i} - 9t\mathbf{j}$, $\qquad \dot{\mathbf{r}} = 6t^2\mathbf{i} - 9\mathbf{j}$.
  When $t = 1$, $\dot{\mathbf{r}} = 6\mathbf{i} - 9\mathbf{j}$.
  Speed $= \sqrt{(6^2 + 9^2)}$
  $\qquad = \sqrt{117}$
  $\qquad = 3\sqrt{13}$ m s^{-1}. $\hspace{3cm}$ <u>Answer **D**</u>

1.5 The position vector $\mathbf{r}$ of a particle at time t is given by

$\mathbf{r} = (2t^2 - 4t + 15)\mathbf{i} + (4t - 7)\mathbf{j}$.

Find the velocity of the particle at time t and the initial velocity (when $t = 0$).

5

At time $t = t_1$, the direction of motion is at right angles to the original direction of motion. Find t_1 and the distance of the particle from its initial position at this time.

- $\mathbf{r} = (2t^2 - 4t + 15)\,\mathbf{i} + (4t - 7)\,\mathbf{j}; \qquad \mathbf{v} = \dot{\mathbf{r}} = (4t - 4)\,\mathbf{i} + 4\mathbf{j}.$
  Initial velocity $\mathbf{v}_0$ (when $t = 0$) is $-4\mathbf{i} + 4\mathbf{j}$.
  Velocity $\mathbf{v}_{t_1}$ at time t_1 is $(4t_1 - 4)\,\mathbf{i} + 4\mathbf{j}$.
  If these velocities are perpendicular, $\mathbf{v}_0 \cdot \mathbf{v}_{t_1} = 0$,
  i.e. $-4(4t_1 - 4) + 4(4) = 0$
  $$-16t_1 + 16 + 16 = 0$$
  so $\qquad\qquad\qquad t_1 = 2.$
  When $t = 0, \qquad \mathbf{r} = 15\mathbf{i} - 7\mathbf{j}.$
  When $t = 2, \qquad \mathbf{r} = 15\mathbf{i} + \mathbf{j}.$
  Thus distance from the initial position is 8 units.

1.6 Two points move in a plane, both starting from the origin, in such a way that, after time t seconds, their position vectors are

$$3(\cos t - 1)\,\mathbf{i} + (4\sin t)\,\mathbf{j} \qquad \text{and} \qquad (3\sin t)\,\mathbf{i} + 4(\cos t - 1)\,\mathbf{j}.$$

Find the values of t when the points are first moving in
(a) opposite directions, (b) the same direction. (L)

- $\mathbf{r}_1 = 3(\cos t - 1)\,\mathbf{i} + 4\sin t\,\mathbf{j}, \qquad \mathbf{r}_2 = 3\sin t\,\mathbf{i} + 4(\cos t - 1)\,\mathbf{j}$
  $\dot{\mathbf{r}}_1 = -3\sin t\,\mathbf{i} + 4\cos t\,\mathbf{j}, \qquad \dot{\mathbf{r}}_2 = 3\cos t\,\mathbf{i} - 4\sin t\,\mathbf{j}.$
  Direction of motion is inclined at an angle $\arctan(\dot{y}/\dot{x})$ to the unit vector $\mathbf{i}$.
  First point moves in a direction inclined at an angle

$$\arctan\left(\frac{4\cos t}{-3\sin t}\right) = \arctan\left(\frac{-4}{3}\cot t\right).$$

Second point moves in a direction inclined at an angle.

$$\arctan\left(\frac{-4\sin t}{3\cos t}\right) = \arctan\left(\frac{-4}{3}\tan t\right).$$

Points are travelling in the same or opposite directions when

$$\frac{-4}{3}\cot t = \frac{-4}{3}\tan t$$

$$\tan^2 t = 1 \quad \Rightarrow \quad \tan t = \pm 1$$

$$\Rightarrow \quad t = \frac{\pi}{4} + \frac{n\pi}{2}.$$

When $t = \dfrac{\pi}{4}, \quad \dot{\mathbf{r}}_1 = -\dfrac{3}{\sqrt{2}}\mathbf{i} + \dfrac{4}{\sqrt{2}}\mathbf{j}, \quad \dot{\mathbf{r}}_2 = \dfrac{3}{\sqrt{2}}\mathbf{i} - \dfrac{4}{\sqrt{2}}\mathbf{j}$, so the points are moving in opposite directions.

When $t = \dfrac{3\pi}{4}, \quad \dot{\mathbf{r}}_1 = -\dfrac{3}{\sqrt{2}}\mathbf{i} - \dfrac{4}{\sqrt{2}}\mathbf{j}, \quad \dot{\mathbf{r}}_2 = -\dfrac{3}{\sqrt{2}}\mathbf{i} - \dfrac{4}{\sqrt{2}}\mathbf{j}$ so the points are moving in the same direction.

Answer (a) Points are first moving in opposite directions when $t = \dfrac{\pi}{4}$.

Answer (b) Points are first moving in the same direction when $t = \dfrac{3\pi}{4}$.

1.7 Two particles P and Q are moving in the x-y plane with constant speeds of 4 m s^{-1} and v m s^{-1} respectively; P moves parallel to the line $4y = 3x$, and Q moves parallel to the line $3y + 4x = 0$ (both in the direction of increasing x). When $t = 0$, P is at the point $(2, 1)$ and Q is at the point $(1, 9)$.

Determine in vector form $a\mathbf{i} + b\mathbf{j}$ (where $\mathbf{i}$ and $\mathbf{j}$ are unit vectors in the directions of x and y respectively):

(a) the velocities of P and Q,   (b) $\overline{OP}, \overline{OQ}$ and $\overline{PQ}$ at time t.

Show that if P and Q collide then $v = 7$, and find the time at which the collision occurs.

- If P moves parallel to the line $4y = 3x$ then $\mathbf{v}_P = 4\lambda\mathbf{i} + 3\lambda\mathbf{j}$.
  But $|\mathbf{v}_P| = 4$, $16\lambda^2 + 9\lambda^2 = 16$
  $$\lambda = \pm\tfrac{4}{5}.$$

Since P moves in the direction of increasing x,

$$\mathbf{v}_P = \frac{16}{5}\mathbf{i} + \frac{12}{5}\mathbf{j}.$$

If Q moves parallel to the line $3y + 4x = 0$ then $\mathbf{v}_Q = 3\mu\mathbf{i} - 4\mu\mathbf{j}$.
But $|\mathbf{v}_Q| = v$, $9\mu^2 + 16\mu^2 = v^2$

$$\mu = \pm\frac{v}{5};$$

$$\mathbf{v}_Q = \pm\left(\frac{3v}{5}\mathbf{i} - \frac{4v}{5}\mathbf{j}\right).$$

Since Q moves in the direction of increasing x,

$$\mathbf{v}_Q = \frac{3v}{5}\mathbf{i} - \frac{4v}{5}\mathbf{j}.$$

By considering initial positions (as shown), at any time t,

$$\overline{OP} = (2\mathbf{i} + \mathbf{j}) + \left(\frac{16}{5}\mathbf{i} + \frac{12}{5}\mathbf{j}\right)t$$

$$= \left(\frac{16}{5}t + 2\right)\mathbf{i} + \left(\frac{12}{5}t + 1\right)\mathbf{j}$$

$$\overline{OQ} = (\mathbf{i} + 9\mathbf{j}) + \left(\frac{3v}{5}\mathbf{i} - \frac{4v}{5}\mathbf{j}\right)t$$

$$= \left(\frac{3v}{5}t + 1\right)\mathbf{i} + \left(9 - \frac{4v}{5}t\right)\mathbf{j}$$

$$\overline{PQ} = \overline{PO} + \overline{OQ} = \left(\frac{3v}{5}t - \frac{16}{5}t - 1\right)\mathbf{i} + \left(8 - \frac{4v}{5}t - \frac{12}{5}t\right)\mathbf{j}.$$

For the particles to collide, $\overline{PQ} = 0$ for some t,

i.e. $\dfrac{3v}{5}t - \dfrac{16}{5}t - 1 = 0, \Rightarrow 3vt - 16t = 5$ (1)

and $8 - \dfrac{4v}{5}t - \dfrac{12}{5}t = 0, \Rightarrow 4vt + 12t = 40 \Rightarrow 3vt + 9t = 30.$ (2)

Subtracting (1) from (2), $\qquad\qquad 25t = 25,$

$$t = 1,$$

$$v = 7.$$

P and Q collide after 1 second, and speed of Q is 7 m s^{-1}.

1.8 A body of mass 20 kg moves under the action of gravity and a force which varies with time given by

$$\mathbf{F} = 80 \sin \pi t\, \mathbf{i} + 80 \cos \pi t\, \mathbf{j} + (240t + 80)\, \mathbf{k}.$$

Horizontal unit vectors in directions east and north are $\mathbf{i}$ and $\mathbf{j}$ respectively, and $\mathbf{k}$ is a unit vector in a vertically upwards direction.

Units of length and time are metres and seconds, and others are SI.
At time $t = 0$ the body starts from rest at O, 5 m above the ground.
Show that, taking g as 10 m s^{-2},

$$\frac{d\mathbf{r}}{dt} = \frac{4}{\pi}(1 - \cos \pi t)\,\mathbf{i} + \frac{4}{\pi}\sin \pi t\,\mathbf{j} + (6t^2 - 6t)\,\mathbf{k},$$

where $\mathbf{r}$ is the position vector of the body relative to O at time t.
Obtain the corresponding expression for $\mathbf{r}$.
Find
(a) the time at which the body is at its lowest point A;
(b) the speed and direction of motion of the body at A;
(c) the distance of A from O.
Show that the body is moving vertically only when $t = 0, 2, 4, 6, \ldots$ seconds.

(SUJB)

• Force due to gravity is $-20g\mathbf{k} = -200\mathbf{k}$ N.
Force acting on the body = $80 \sin \pi t\,\mathbf{i} + 80 \cos \pi t\,\mathbf{j} + (240t - 120)\,\mathbf{k}$.
But $\mathbf{F} = m\mathbf{a}$ $\Rightarrow$ $\mathbf{a} = \ddot{\mathbf{r}} = 4 \sin \pi t\,\mathbf{i} + 4 \cos \pi t\,\mathbf{j} + (12t - 6)\,\mathbf{k}$,

$$\frac{d\mathbf{r}}{dt} = \int \frac{d^2\mathbf{r}}{dt^2}\, dt = \left(-\frac{4}{\pi}\cos \pi t + c_1\right)\mathbf{i} + \left(\frac{4}{\pi}\sin \pi t + c_2\right)\mathbf{j} + (6t^2 - 6t + c_3)\mathbf{k}$$

When $t = 0$, $\dfrac{d\mathbf{r}}{dt} = 0$, $\Rightarrow$ $c_1 = \dfrac{4}{\pi}$, $c_2 = 0$, $c_3 = 0$.

$$\mathbf{v} = \frac{d\mathbf{r}}{dt} = \frac{4}{\pi}(1 - \cos \pi t)\,\mathbf{i} + \frac{4}{\pi}\sin \pi t\,\mathbf{j} + (6t^2 - 6t)\,\mathbf{k},$$

$$\mathbf{r} = \int \frac{d\mathbf{r}}{dt}\, dt = \frac{4}{\pi}\left(t - \frac{1}{\pi}\sin \pi t + c_4\right)\mathbf{i} + \left(-\frac{4}{\pi^2}\cos \pi t + c_5\right)\mathbf{j} + (2t^3 - 3t^2 + c_6)\,\mathbf{k}.$$

When $t = 0$, $\mathbf{r} = 0$ relative to O, $c_4 = 0$, $c_5 = \dfrac{4}{\pi^2}$, $c_6 = 0$,

$$\mathbf{r} = \frac{4}{\pi^2}(\pi t - \sin \pi t)\,\mathbf{i} + \frac{4}{\pi^2}(1 - \cos \pi t)\,\mathbf{j} + (2t^3 - 3t^2)\,\mathbf{k}.$$

When the body is at its lowest point, vertical velocity = 0, _Why?_
i.e. $6t^2 - 6t = 0$ $\Rightarrow$ $t = 0$ or $t = 1$.   _Why not $2t^3 - 3t^2 = 0$_
                                                                        m whi case $2t - 3 = 0$
$t = 0$ is the starting time.   _$t = \frac{3}{2}$._
When $t = 1$, vertical acceleration = $6\mathbf{k}$, i.e. acceleration is vertically upwards.
(a) Body reaches its lowest point A after 1 second.

(b) $\dot{\mathbf{r}}(1) = \dfrac{8}{\pi}\mathbf{i}$,   i.e. at A velocity is $\dfrac{8}{\pi}$ m s^{-1} due east.

(c) $\mathbf{r}(1) = \dfrac{4}{\pi}\mathbf{i} + \dfrac{8}{\pi^2}\mathbf{j} - 1\mathbf{k}$.

The distance of A from O is $\sqrt{\left(\dfrac{16}{\pi^2} + \dfrac{64}{\pi^4} + 1\right)} = 1 + \dfrac{8}{\pi^2}$.

$\sqrt{\dfrac{16\pi^2 + 64 + \pi^4}{\pi^4}} = \dfrac{1}{\pi^2}\sqrt{(\pi^2 + 8)^2} = \dfrac{\pi^2 + 8}{\pi^2}$

Component of velocity in $\mathbf{i}$ direction is zero when $\cos \pi t = 1$,

i.e. $$\pi t = 2n\pi, \qquad t = 2n. \tag{1}$$

Component of velocity in $\mathbf{j}$ direction is zero when $\sin \pi t = 0$,

i.e. $$\pi t = n\pi, \qquad t = n. \tag{2}$$

(1) and (2) are both satisfied when $t = 2n$.
Body is moving vertically when $t = 0, 2, 4, 6 \ldots$ seconds.

1.9   A particle A is at rest on a smooth horizontal table at a point whose position vector relative to a fixed origin O is $(-\mathbf{i} + 2\mathbf{j})$ m, where $\mathbf{i}$ and $\mathbf{j}$ are mutually perpendicular unit vectors in the plane of the table. A point B of the table has position vector $(2\mathbf{i} + \mathbf{j})$ m with respect to O. At time $t = 0$ the particle A is projected along the table with velocity $(6\mathbf{i} + 3\mathbf{j})$ m s^{-1}. Determine the vectors $\overline{OA}$ and $\overline{BA}$ at time t.

Find the time at which the distance BA is equal to 5 m, and for this time find also:
(a) the unit vector $\mathbf{c}$ along $\overline{BA}$,
(b) the unit vector $\mathbf{d}$ perpendicular to $\overline{BA}$.

Hence express the velocity of A in the form $p\mathbf{c} + q\mathbf{d}$, where p and q are scalars, which are to be determined. **(AEB 1983)**

● Let $\mathbf{r}_A$ be the position vector of A in metres.

$$\dot{\mathbf{r}}_A = 6\mathbf{i} + 3\mathbf{j} \quad \Rightarrow \quad \mathbf{r}_A = (-\mathbf{i} + 2\mathbf{j}) + (6\mathbf{i} + 3\mathbf{j})\,t,$$

$$\therefore \quad \mathbf{r}_A = \overline{OA} = (6t - 1)\mathbf{i} + (3t + 2)\mathbf{j}.$$

$$\overline{BA} = \overline{BO} + \overline{OA}$$

$$= -(2\mathbf{i} + \mathbf{j}) + (6t - 1)\mathbf{i} + (3t + 2)\mathbf{j}$$

$$= (6t - 3)\mathbf{i} + (3t + 1)\mathbf{j}.$$

$$(BA)^2 = (6t - 3)^2 + (3t + 1)^2 = 36t^2 - 36t + 9 + 9t^2 + 6t + 1$$

$$= 45t^2 - 30t + 10.$$

When $BA = 5$, $$25 = 45t^2 - 30t + 10,$$

$$\therefore 15\,(3t^2 - 2t - 1) = 0,$$

$$(3t + 1)\,(t - 1) = 0,$$

$$(t = -\tfrac{1}{3}) \text{ or } t = 1 \text{ seconds.}$$

When $t = 1$, $\quad \overline{BA} = 3\mathbf{i} + 4\mathbf{j}$.
(a) Unit vector along $\overline{BA}$, $\mathbf{c} = \tfrac{1}{5}\,(3\mathbf{i} + 4\mathbf{j})$.
(b) Unit vector perpendicular to $\overline{BA}$, $\quad \mathbf{d} = \tfrac{1}{5}(4\mathbf{i} - 3\mathbf{j})$.*
Let the velocity of $A = (p\mathbf{c} + q\mathbf{d})$ m s^{-1}.
Then $\dot{\mathbf{r}}_A = p\mathbf{c} + q\mathbf{d} = \dfrac{p}{5}\,(3\mathbf{i} + 4\mathbf{j}) + \dfrac{q}{5}\,(4\mathbf{i} - 3\mathbf{j})$.

But $\dot{\mathbf{r}}_A = 6\mathbf{i} + 3\mathbf{j}$

$$\Rightarrow \quad \frac{3p}{5} + \frac{4q}{5} = 6 \quad \Rightarrow \quad 3p + 4q = 30, \tag{1}$$

$$\frac{4p}{5} - \frac{3q}{5} = 3 \quad \Rightarrow \quad 4p - 3q = 15. \tag{2}$$

$$3(1) + 4(2) \Rightarrow 25p = 150; \qquad p = 6, \quad q = 3.$$

Hence velocity of A is $(6\mathbf{c} + 3\mathbf{d})$ m s^{-1}.

*$\mathbf{d}$ could equally well be $-\frac{1}{5}(4\mathbf{i} - 3\mathbf{j})$, giving $q = -3$.

1.10 A particle of unit mass is acted on by three forces with the following magnitudes and directions:

$\sqrt{3}$ N due north,

3 N in the direction S. 30° E.,

$2\sqrt{3}$ N in the direction S. 60° W.

(a) Find the magnitude and direction of the resultant force.

An additional force now acts on the particle, so that it has an acceleration of 3 m s^{-2} due east.

(b) Find the magnitude and direction of the extra force.

● (a) Let $\mathbf{i}$ and $\mathbf{j}$ be unit vectors due east and south respectively.

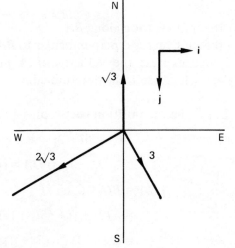

Let $\mathbf{R}$ be the resultant force.

$$\mathbf{R} = (0\mathbf{i} - \sqrt{3}\mathbf{j}) + [3(\sin 30°)\mathbf{i} + 3(\cos 30°)\mathbf{j}] + [-2\sqrt{3}(\sin 60°)\mathbf{i} + 2\sqrt{3}(\cos 60°)\mathbf{j}]$$

$$= \left(\frac{3}{2} - 3\right)\mathbf{i} + \left(-\sqrt{3} + \frac{3\sqrt{3}}{2} + \sqrt{3}\right)\mathbf{j}$$

$$= -\frac{3}{2}\mathbf{i} + \frac{3\sqrt{3}}{2}\mathbf{j}.$$

Resultant has magnitude $\sqrt{\left[\left(-\frac{3}{2}\right)^2 + \left(\frac{3\sqrt{3}}{2}\right)^2\right]}$

$$= \sqrt{\left[\frac{9}{4} + \frac{27}{4}\right]}$$

$$= 3 \text{ N.}$$

Direction is S. $\arctan\left(\dfrac{3/2}{(3\sqrt{3})/2}\right)$ W., i.e. S. 30° W.

(b) If the force required to give the particle an acceleration of 3 m s^{-2} due east is $x\mathbf{i} + y\mathbf{j}$ then

$$\left(-\frac{3}{2} + x\right)\mathbf{i} + \left(\frac{3\sqrt{3}}{2} + y\right)\mathbf{j} = 3m\mathbf{i},$$

where m is the mass of the particle.

But $m = 1 \Rightarrow -1.5 + x = 3, \Rightarrow x = 4.5.$

10

And $y + \dfrac{3\sqrt{3}}{2} = 0,\quad \Rightarrow \quad y = -\dfrac{3\sqrt{3}}{2}.$

Magnitude of the extra force $= \sqrt{\left[\left(\dfrac{9}{2}\right)^2 + \left(-\dfrac{3\sqrt{3}}{2}\right)^2\right]} = \sqrt{\left(\dfrac{81}{4} + \dfrac{27}{4}\right)} = \sqrt{27}$

$$= 3\sqrt{3}\ \text{N}.$$

Direction is N. $\arctan\left(\dfrac{9/2}{(3\sqrt{3})/2}\right)$ E., i.e. N. $\arctan\sqrt{3}$ E., i.e. N. $60°$ E.

Alternative method for (b)
From the diagram the triangle of forces is isosceles.
Extra force required $= 2\,(3\sin 60°)$ N. $= 3\sqrt{3}$ N. in direction N. $60°$ E.

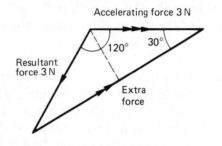

1.3 Exercises

1.1 A point P has position vector $\mathbf{r}$ with respect to the origin O at time t, where $\mathbf{r} = (a\cos pt)\,\mathbf{i} + (b\sin pt)\,\mathbf{j}$, where a, b, and p are constants.

Show that $\dfrac{d^2\mathbf{r}}{dt^2} + p^2\,\mathbf{r} = 0$ throughout the motion.

1.2 The position vector $\mathbf{r}$ of a particle at time t is

$$\mathbf{r} = (a\cos\omega t)\,\mathbf{i} + (b\sin\omega t)\,\mathbf{j} \qquad (a \neq b),$$

where a, b and ω are constants and $\mathbf{i}$, $\mathbf{j}$ are perpendicular unit vectors.
Find the velocity and acceleration vectors $\mathbf{v}$ and $\mathbf{f}$.
Show that $\omega^2\mathbf{v}\cdot\mathbf{v} + \mathbf{f}\cdot\mathbf{f} = \omega^4\,(a^2 + b^2)$ and find the times at which vectors $\mathbf{v}$ and $\mathbf{f}$ are perpendicular.

1.3 A particle of mass 500 g moves under a force $\mathbf{F}$ so that its position vector after t seconds is given by

$$\mathbf{r} = (1 + t + 5t^2 - 3t^3)\,\mathbf{i} + (3 - 4t)\,\mathbf{j}.$$

Find $\mathbf{F}$ and the magnitude of the force after 3 seconds.

1.4 Vectors $\overline{OA}$ and $\overline{OB}$ are given by $\quad \overline{OA} = 3s\mathbf{i} - 4t\mathbf{j},\quad \overline{OB} = 4t\mathbf{i} + 3s\mathbf{j}.$
(a) Find the length of AB in terms of s and t, and show that the angle AOB is $90°$.
(b) Find the shortest distance OC from O to AB and the position vector of C relative to O.

1.5 At time t two points P and Q have position vectors $\mathbf{p}$ and $\mathbf{q}$ respectively, where

$$\mathbf{p} = 2a\mathbf{i} + (a\cos\omega t)\mathbf{j} + (a\sin\omega t)\mathbf{k}, \qquad \mathbf{q} = (a\sin\omega t)\mathbf{i} - (a\cos\omega t)\mathbf{j} + 3a\mathbf{k},$$

and a and ω are constants. Find $\mathbf{r}$, the position vector of P relative to Q and $\mathbf{v}$, the velocity of P relative to Q. Find also the values of t for which $\mathbf{r}$ and $\mathbf{v}$ are perpendicular.

Determine the smallest and greatest distances between P and Q. (L)

1.6 The position vector of a particle P with respect to a fixed origin O is $-(\cos 2t)\mathbf{i} + (\sin 2t)\mathbf{j} + t\mathbf{k}$ m, where t is the time in seconds $(t \geqslant 0)$.

The position vector of particle Q with respect to O is $\mathrm{e}^t\mathbf{i} - \mathrm{e}^t\mathbf{j} + \mathbf{k}$ m. Find the first two values of t for which the velocities of P and Q are perpendicular and show that, for the first of these values, the accelerations of the particles are parallel.

1.7 A particle of unit mass moves under the influence of a force such that when the position vector of the particle is $\overline{OP} = x\mathbf{i} + y\mathbf{j}$, it experiences a force $\mathbf{F} = -n^2 x\mathbf{i} - 4n^2 y\mathbf{j}$, where n is a constant.

At time $t = 0$, the particle has position vector $a\mathbf{i}$ and velocity vector $na\mathbf{j}$. Find x and y in terms of n and t, and find the locus of P.

(Standard solutions of differential equations may be stated.)

1.8   In the triangle ABC, D is a point on BC such that $BD : DC = \alpha : \beta$.

Show that $(\alpha + \beta)\,\overline{AD} = \alpha\,\overline{AC} + \beta\,\overline{AB}$.

The non-collinear points A, B and C have position vectors $\mathbf{a}$, $\mathbf{b}$ and $\mathbf{c}$ respectively with respect to any origin O. The point E on AC is such that $AE : EC = 2 : 3$, and the point D on CB is such that $CD : DB = 4 : 1$.

Show that $\overline{AD} = -\mathbf{a} + \frac{4}{5}\mathbf{b} + \frac{1}{5}\mathbf{c}$, and find a similar expression for $\overline{BE}$.

The lines AD and BE intersect at X.

By considering $\overline{AX}$ as part of $\overline{AD}$ and $\overline{BX}$ as part of $\overline{BE}$, find the position vector of X in terms of $\mathbf{a}$, $\mathbf{b}$ and $\mathbf{c}$, and find the ratio $BX : XE$.

1.4 Brief Solutions to Exercises

1.1 $\dot{\mathbf{r}} = \dfrac{\mathrm{d}\mathbf{r}}{\mathrm{d}t} = -(ap\sin pt)\mathbf{i} + (bp\cos pt)\mathbf{j};$

$$\ddot{\mathbf{r}} = \dfrac{\mathrm{d}^2\mathbf{r}}{\mathrm{d}t^2} = -(ap^2\cos pt)\mathbf{i} - (bp^2\sin pt)\mathbf{j} = -p^2\mathbf{r}.$$

Hence $\dfrac{\mathrm{d}^2\mathbf{r}}{\mathrm{d}t^2} + p^2\mathbf{r} = 0.$

1.2 $\mathbf{r} = (a\cos\omega t)\mathbf{i} + (b\sin\omega t)\mathbf{j}.$
$\mathbf{v} = \dot{\mathbf{r}} = -(a\omega\sin\omega t)\mathbf{i} + (b\omega\cos\omega t)\mathbf{j}. \qquad \mathbf{f} = \ddot{\mathbf{r}} = -(a\omega^2\cos\omega t)\mathbf{i} - (b\omega^2\sin\omega t)\mathbf{j}.$
$\mathbf{v} \cdot \mathbf{v} = a^2\omega^2\sin^2\omega t + b^2\omega^2\cos^2\omega t;$
$\mathbf{f} \cdot \mathbf{f} = a^2\omega^4\cos^2\omega t + b^2\omega^4\sin^2\omega t;$
$\omega^2\mathbf{v} \cdot \mathbf{v} + \mathbf{f} \cdot \mathbf{f} = a^2\omega^4(\sin^2\omega t + \cos^2\omega t) + b^2\omega^4(\cos^2\omega t + \sin^2\omega t),$
$\omega^2\mathbf{v} \cdot \mathbf{v} + \mathbf{f} \cdot \mathbf{f} = \omega^4(a^2 + b^2).$

$\mathbf{f} \cdot \mathbf{v} = a^2 \omega^3 \sin \omega t \cos \omega t - b^2 \omega^3 \sin \omega t \cos \omega t$

$\qquad = \omega^3 \sin \omega t \cos \omega t \, (a^2 - b^2);$

$\mathbf{f} \cdot \mathbf{v} = 0$ when $\sin \omega t \cos \omega t = 0.$

i.e. $\omega t = \dfrac{n\pi}{2}, \qquad t = \dfrac{n\pi}{2\omega}.$

1.3 $\quad \mathbf{r} = (1 + t + 5t^2 - 3t^3)\,\mathbf{i} + (3 - 4t)\,\mathbf{j};$

$\dot{\mathbf{r}} = (1 + 10t - 9t^2)\,\mathbf{i} - 4\mathbf{j};$

$\ddot{\mathbf{r}} = (10 - 18t)\,\mathbf{i}.$

Acceleration after t seconds is $(10 - 18t)\,\mathbf{i}.$

But force = mass $\times$ acceleration:

$$\mathbf{F} = 0.5\,(10 - 18t)\,\mathbf{i} = (5 - 9t)\,\mathbf{i}.$$

When $t = 3$, $\mathbf{F} = -22\mathbf{i}$, i.e. magnitude of the force is 22 N.

1.4 (a) $\overline{AB} = -3s\mathbf{i} + 4t\mathbf{j} + 4t\mathbf{i} + 3s\mathbf{j} = (4t - 3s)\,\mathbf{i} + (4t + 3s)\,\mathbf{j}.$ $\qquad |\overline{AB}| = \sqrt{(32t^2 + 18s^2)}.$

$\overline{OA} \cdot \overline{OB} = 12st - 12st = 0 \;\Rightarrow\; \cos A\hat{O}B = 0 \;\Rightarrow\; A\hat{O}B = \dfrac{\pi}{2}.$

(b) $|\overline{OA}| = \sqrt{(9s^2 + 16t^2)}, \qquad |\overline{OB}| = \sqrt{(16t^2 + 9s^2)}, \qquad |\overline{OA}| = |\overline{OB}|$

$\Rightarrow \quad OAB$ is a right-angled isosceles triangle and the least distance from O to AB is

when C is the mid-point of $AB \;\Rightarrow\; \overline{OC} = \dfrac{3s + 4t}{2}\,\mathbf{i} + \dfrac{3s - 4t}{2}\,\mathbf{j}.$

$$|\overline{OC}| = \frac{1}{2}\sqrt{(32t^2 + 18s^2)}.$$

Alternatively $\quad OC = \frac{1}{2}AB \quad$ from a right-angled isosceles triangle.

1.5 $\quad \mathbf{r} = \overline{QP} = \overline{QO} + \overline{OP} = (2a - a \sin \omega t)\,\mathbf{i} + (2a \cos \omega t)\,\mathbf{j} + (a \sin \omega t - 3a)\,\mathbf{k}.$

$\mathbf{v} = \dot{\mathbf{r}} = -(a\omega \cos \omega t)\,\mathbf{i} - (2a\omega \sin \omega t)\,\mathbf{j} + (a\omega \cos \omega t)\,\mathbf{k}.$

When $\mathbf{r}$ and $\mathbf{v}$ are perpendicular, $\mathbf{r} \cdot \mathbf{v} = 0.$

$-5 \cos \omega t - 2 \sin \omega t \cos \omega t = 0, \;\Rightarrow\; \cos \omega t = 0,$

$$t = \frac{\pi}{2\omega}(2n - 1).$$

Distance PQ is max. or min. when $\mathbf{r} \cdot \mathbf{v} = 0$, i.e. when $t = \dfrac{\pi}{2\omega}(2n - 1);$

when $n = 1$, $\quad |\mathbf{r}| = a\sqrt{5}; \qquad$ when $n = 2$, $\quad |\mathbf{r}| = 5a,$

$\Rightarrow$ greatest and least distances PQ are $5a$ and $a\sqrt{5}.$

1.6 For P, $\quad \mathbf{r}_P = -(\cos 2t)\,\mathbf{i} + (\sin 2t)\,\mathbf{j} + t\mathbf{k},$

$\qquad\qquad \dot{\mathbf{r}}_P = (2 \sin 2t)\,\mathbf{i} + (2 \cos 2t)\,\mathbf{j} + \mathbf{k},$

$\qquad\qquad \ddot{\mathbf{r}}_P = (4 \cos 2t)\,\mathbf{i} - (4 \sin 2t)\,\mathbf{j}.$

For Q, $\quad \mathbf{r}_Q = e^t\mathbf{i} - e^t\mathbf{j} + \mathbf{k},$

$\qquad\qquad \dot{\mathbf{r}}_Q = e^t\mathbf{i} - e^t\mathbf{j},$

$\qquad\qquad \ddot{\mathbf{r}}_Q = e^t\mathbf{i} - e^t\mathbf{j}.$

When velocities are perpendicular, $\quad \dot{\mathbf{r}}_P \cdot \dot{\mathbf{r}}_Q = 0,$

i.e. $\quad e^t (\sin 2t - \cos 2t) = 0 \qquad (e^t \ne 0),$

$$\sin 2t = \cos 2t, \quad t = \frac{\pi}{8}, \; \frac{5\pi}{8}.$$

When $t = \dfrac{\pi}{8}$, $\quad \ddot{r}_P = (2\sqrt{2})\,\mathbf{i} - (2\sqrt{2})\,\mathbf{j}$

$$= 2\sqrt{2}\,(\mathbf{i} - \mathbf{j});$$
$$\ddot{r}_Q = e^{\pi/8}\,\mathbf{i} - e^{\pi/8}\,\mathbf{j}$$
$$= e^{\pi/8}\,(\mathbf{i} - \mathbf{j});$$

accelerations are parallel.

1.7 Using $\mathbf{F} = m\mathbf{a}$, $\qquad \ddot{x} = -n^2 x$, $\qquad \ddot{y} = -4n^2 y$.

Solutions: $x = A\cos nt + B\sin nt$, $\qquad y = C\cos 2nt + D\sin 2nt$;

$\qquad\qquad \dot{x} = Bn\cos nt - An\sin nt$, $\qquad \dot{y} = 2Dn\cos 2nt - 2Cn\sin 2nt$.

Substituting initial conditions gives $\quad A = a$, $\quad B = 0$, $\quad C = 0$, $\quad D = \dfrac{a}{2}$.

$$x = a\cos nt, \quad y = \frac{a}{2}\sin 2nt.$$

The locus of P is obtained by eliminating t from these two equations.

$a^2 y^2 = x^2\,(a^2 - x^2)$.

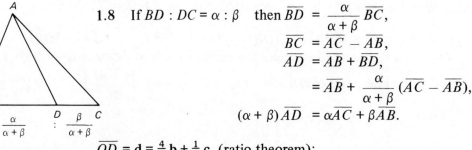

1.8 If $BD : DC = \alpha : \beta$ $\quad$ then $\overline{BD} = \dfrac{\alpha}{\alpha + \beta}\,\overline{BC}$,

$$\overline{BC} = \overline{AC} - \overline{AB},$$
$$\overline{AD} = \overline{AB} + \overline{BD},$$
$$= \overline{AB} + \frac{\alpha}{\alpha + \beta}\,(\overline{AC} - \overline{AB}),$$
$$(\alpha + \beta)\,\overline{AD} = \alpha\overline{AC} + \beta\overline{AB}.$$

$\overline{OD} = \mathbf{d} = \frac{4}{5}\mathbf{b} + \frac{1}{5}\mathbf{c}$ (ratio theorem);

$\overline{AD} = \overline{AO} + \overline{OD} = -\mathbf{a} + \frac{4}{5}\mathbf{b} + \frac{1}{5}\mathbf{c}$.

Similarly, $\overline{BE} = -\mathbf{b} + \frac{3}{5}\mathbf{a} + \frac{2}{5}\mathbf{c}$.

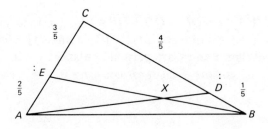

If $\overline{AX} = s\overline{AD}$, and $\overline{BX} = t\overline{BE}$, then $\overline{AX} + \overline{XB} = s\overline{AD} - t\overline{BE}$.

$\quad$ But $\overline{AX} + \overline{XB} = \overline{AB} = \mathbf{b} - \mathbf{a}$.

Thus $\mathbf{a}\,(-s - \frac{3}{5}t) + \mathbf{b}\,(\frac{4}{5}s + t) + \mathbf{c}\,(\frac{1}{5}s - \frac{2}{5}t) = \mathbf{b} - \mathbf{a}$ $\Rightarrow$ $s = \frac{10}{13}$, $t = \frac{5}{13}$.

$\overline{OX} = \overline{OA} + \overline{AX} = \mathbf{a} - \frac{10}{13}\mathbf{a} + \frac{8}{13}\mathbf{b} + \frac{2}{13}\mathbf{c} = \frac{1}{13}(3\mathbf{a} + 8\mathbf{b} + 2\mathbf{c})$;

$BX : XE$ is $5 : 8$.

2 Relative Motion

Relative velocity and displacement. Graphical methods.

2.1 Fact Sheet

(a) Relative Velocity

If $\mathbf{r}_A$ and $\mathbf{r}_B$ are the position vectors of A and B then $\dot{\mathbf{r}}_A = \mathbf{v}_A$ and $\dot{\mathbf{r}}_B = \mathbf{v}_B$ are the velocities of A and B.

The velocity of A relative to B is given by $_A\mathbf{v}_B = \mathbf{v}_A - \mathbf{v}_B$.

The velocity of B relative to A is given by $_B\mathbf{v}_A = \mathbf{v}_B - \mathbf{v}_A$.

Hence, $_A\mathbf{v}_B = - {_B}\mathbf{v}_A$.

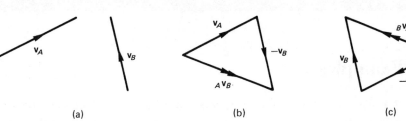

For three moving points, $_A\mathbf{v}_B = {_A}\mathbf{v}_C - {_B}\mathbf{v}_C$. or $_A\mathbf{v}_B = {_A}\mathbf{v}_C + {_C}\mathbf{v}_B$.

(b) Displacement

$\mathbf{r}_A - \mathbf{r}_B = \overline{BA}$, the displacement of A relative to B.

$\mathbf{r}_B - \mathbf{r}_A = \overline{AB}$, the displacement of B relative to A.

Distance between A and B is $|\overline{BA}| = |\overline{AB}|$.

Least distance occurs when $\overline{AB} \cdot {_A}\mathbf{v}_B = 0$ (or when $|\overline{AB}|^2$ is a minimum).

Collision will occur if $_A\mathbf{v}_B = k\overline{AB}$, provided $_A\mathbf{v}_B$ is a constant vector.

(c) Graphical Method

This can be two distinct constructions, one for velocities and one for displacements, or one can be superimposed upon the other (usually more accurate and quicker).

  (i) Choose scales for distances (to be used for all distances) and for velocities (to be used for all velocities).

  (ii) Mark original positions of A and B on a diagram (L and M).

  (iii) To find $_A\mathbf{v}_B$, construct the velocity vector triangle LCE. $_A\mathbf{v}_B = \overline{LE}$ in both magnitude and direction.

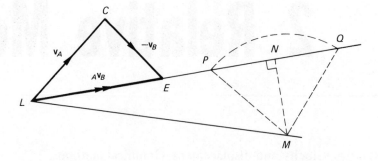

Least distance between particles is given by MN, where N is the foot of the perpendicular from M to LE (produced if necessary).

To find the time needed to reach any particular distance between the particles, find the one/two points P and Q on the line LE (produced if necessary), where $MP = MQ =$ required distance.

Time taken: $\dfrac{LP}{|\,_A\mathbf{v}_B\,|}$ or $\dfrac{LQ}{|\,_A\mathbf{v}_B\,|}$.

Collision occurs only if $MN = 0$, i.e. if LE is along LM. For collision, the velocity of A relative to B must be in the direction LM, the initial positions of A and B.

2.2 Worked Examples

2.1 A ship P is moving due west at 12 km h^{-1}. The velocity of a second ship Q relative to P is 15 km h^{-1} in a direction $30°$ west of south. Find the velocity of ship Q.

- Let the velocities of P and Q be $\mathbf{v}_P$ and $\mathbf{v}_Q$ respectively, and the velocity of Q relative to P be $_Q\mathbf{v}_P$.
  Then $_Q\mathbf{v}_P = \mathbf{v}_Q - \mathbf{v}_P \Rightarrow \mathbf{v}_Q = {_Q\mathbf{v}_P} + \mathbf{v}_P$.
  Method 1: By vectors
  Let $\mathbf{w}$ and $\mathbf{s}$ be unit vectors due west and south respectively.

  $\mathbf{v}_P = 12\mathbf{w}, \quad _Q\mathbf{v}_P = 15 \sin 30\mathbf{w} + 15 \cos 30\mathbf{s} = \dfrac{15}{2}\mathbf{w} + \dfrac{15}{2}\sqrt{3}\mathbf{s}.$

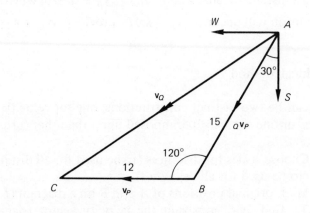

Hence $\mathbf{v}_Q = \dfrac{39}{2}\mathbf{w} + \dfrac{15}{2}\sqrt{3}\mathbf{s}.$

Speed of $Q = |v_Q| = \frac{1}{2}\sqrt{[(39)^2 + (15\sqrt{3})^2]}$
$= 23.4$ km h^{-1}.

Direction of motion of Q is S. $\alpha°$ W. where $\tan \alpha = 39/(15\sqrt{3})$. Direction is S. 56.3° W.

Method 2: Using trigonometry on triangle ABC
By the cosine rule:
$AC^2 = 144 + 225 - 2(12)(15) \cos 120$
$= 549;$
AC = the speed of $Q = 23.4$ km h^{-1}.
By the sine rule, $(\sin A)/12 = (\sin 120)/23.4$,
$A = 26.3°$.
Hence the direction of motion of Q is S. 56.3° W.

Method 3: If acceptable, by construction.

2.2   In water which has a steady current from the east of 5 knots, a ship sails at constant speed, relative to the water, of 15 knots. The ship moves in a straight line from port P to port Q which is 100 nautical miles south west of P.
    Denoting the true speed of the ship as x knots, show that

$$x^2 - 5\sqrt{2}x - 200 = 0.$$

Hence find x in surd form.
    After 12 hours in port Q the ship returns to P, experiencing the same current. Find the total time for which the ship was away from P.

● Let ABC represent the triangle of velocities.
    $AB = 5$, $BC = 15$, $A = 45°$, $AC = x$.

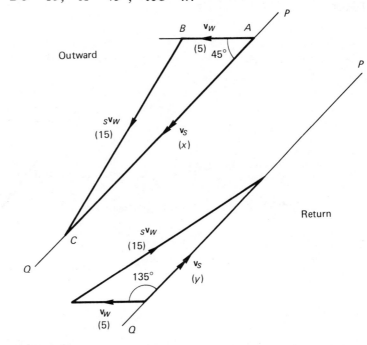

By the cosine rule,

$$15^2 = 5^2 + x^2 - 10x \cos 45,$$

i.e. $x^2 - 5\sqrt{2}x - 200 = 0$,

$$x = \frac{5\sqrt{2} \pm \sqrt{(50 + 800)}}{2} = \frac{5\sqrt{2} \pm \sqrt{850}}{2}.$$

Since x is positive, $x = \dfrac{5\sqrt{2} + \sqrt{850}}{2}$ $(= 18.11)$.

17

Return journey

If the true speed is now y knots,

$$15^2 = 5^2 + y^2 - 10y \cos 135,$$

i.e. $\quad y^2 + 5\sqrt{2}y - 200 = 0.$

$$y = \frac{-5\sqrt{2} + \sqrt{850}}{2} \quad (= 11.04).$$

Total time taken $= 12 + \dfrac{100}{x} + \dfrac{100}{y} = 26.58$ h or 26 h 35 min.

2.3 Two ships A and B are sailing with constant speeds of 15 knots due east and 20 knots due north respectively. Their courses intersect at a point C. At noon B is at C and A is 7.5 nautical miles due west of C. Find:
(a) the velocity of ship B relative to ship A;
(b) the least distance between the ships;
(c) the course which A should have taken at noon to minimise the distance between the ships, assuming that A continued at the same speed;
(d) the time A would have taken to reach the position of minimum distance.

- Let $\quad CX$ represent the velocity of B,
  $\quad\quad\quad YX$ represent the velocity of A,
  and $\quad CY$ represent the velocity of B relative to A.

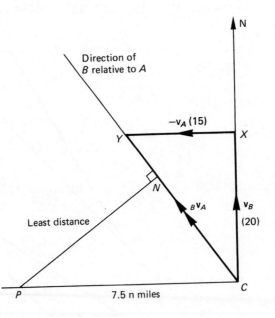

From triangle CXY, $CY = 25$, $\quad C = \arctan(15/20) = 36.9°$.
(a) Velocity of B relative to A is 25 knots, N. 36.9° W.
(b) If A is at P at noon, then, from the diagram, the least distance between the ships is PN, the perpendicular distance from P to CY (produced if necessary). $PN = 7.5 \sin 53.1° = 6$ n m.
(c) For the distance between the ships to be a minimum the velocity of B relative to A must make the least possible angle with CP. Therefore the velocity of A must be perpendicular to $_B\mathbf{v}_A$.

18

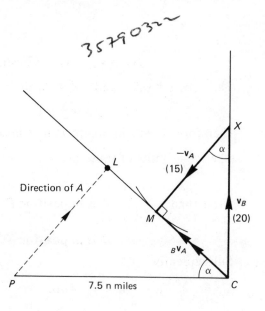

From the diagram $\cos\alpha = 15/20, \quad \alpha = 41.4°$.
$|_B v_A| = CM = 20 \sin\alpha = 13.2$ knots.
Course set by A should have been N. $41.4°$ E.

(d) CL represents the distance travelled by B relative to A to reach the closest point.
$CL = 7.5 \cos\alpha = 5.625$ n m.
Time taken $= 5.625/13.2 = 0.426$ h.
Hence A would have reached this point at 25.6 min past noon.

Alternatively: Construct the figures accurately and take the necessary measurements from the construction.

2.4 A and B move in a horizontal plane with constant speeds $2u$ and u respectively. At time $t = 0$, B is distant $2a$ due east of A and moving northwards. Show that A must move in the direction N. $60°$ E. in order to intercept B and find the time that it would take to do so. If, after half this time has elapsed, A's speed drops to u without changing direction, find how close A comes to B and the distance that A has moved from its original position when they are closest.

(SUJB)

● Initially A is at point X and B is at point W. In order to intercept B the velocity of A relative to B must be due east.

Let **e** and **n** be unit vectors due east and north respectively, and let A head in the direction N. $\alpha°$ E.:

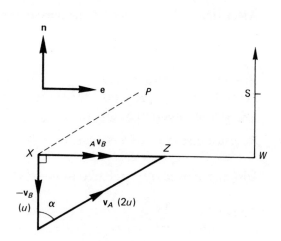

19

$$_A\mathbf{v}_B = \mathbf{v}_A - \mathbf{v}_B = (2u\sin\alpha\,\mathbf{e} + 2u\cos\alpha\,\mathbf{n}) - u\mathbf{n} = k\mathbf{e},$$

i.e. $\cos\alpha = 0.5$ and $k = 2u\sin\alpha$, $\alpha = 60°$, $k = u\sqrt{3}$,

$$\mathbf{v}_A = \sqrt{3}u\mathbf{e} + u\mathbf{n}$$

Therefore A will intercept B if it heads in the direction N. $60°$ E. and it will

take $\dfrac{2a}{u\sqrt{3}}$ units of time.

After time $\dfrac{a}{u\sqrt{3}}$, A is in position P $a\mathbf{e} + \dfrac{a\,\mathbf{n}}{\sqrt{3}}$ relative to X,

B is in position S $2a\mathbf{e} + \dfrac{a}{\sqrt{3}}\,\mathbf{n}$ relative to X.

Subsequently

$$\mathbf{v}'_A = \frac{u\sqrt{3}}{2}\,\mathbf{e} + \frac{u}{2}\,\mathbf{n}, \quad \mathbf{v}_B = u\mathbf{n}.$$

Positions of A and B after a further time t are

$$\left(\frac{u\sqrt{3}}{2}t + a\right)\mathbf{e} + \left(\frac{u}{2}t + \frac{a}{\sqrt{3}}\right)\mathbf{n} \quad \text{and} \quad 2a\mathbf{e} + \left(ut + \frac{a}{\sqrt{3}}\right)\mathbf{n} \quad \text{respectively.}$$

Displacement vector $\overline{AB}_t = \left(a - \dfrac{u\sqrt{3}t}{2}\right)\mathbf{e} + \dfrac{ut}{2}\,\mathbf{n}.$

$$(\text{Distance } AB)^2 = \left(a - \frac{u\sqrt{3}t}{2}\right)^2 + \left(\frac{ut}{2}\right)^2 = a^2 - \sqrt{3}aut + u^2t^2.$$

Either by differentiation or by the property of a quadratic having a turning

point at $\dfrac{-b}{2a}$, minimum value occurs when $ut = \dfrac{a\sqrt{3}}{2}$.

Then the least distance $= \sqrt{\left(a^2 - \dfrac{3}{2}a^2 + \dfrac{3}{4}a^2\right)} = \dfrac{a}{2}$.

At this instant the position of A is:

$$\left(\frac{3a}{4} + a\right)\mathbf{e} + \left(\frac{a\sqrt{3}}{4} + \frac{a}{\sqrt{3}}\right)\mathbf{n} = \frac{7a}{4}\mathbf{e} + \frac{7a}{4\sqrt{3}}\mathbf{n}.$$

Distance from the original position $= \dfrac{7a}{4}\sqrt{\left[1^2 + \left(\dfrac{1}{\sqrt{3}}\right)^2\right]}$

$$= \frac{7a}{2\sqrt{3}}.$$

Alternative method for second part:

After time $\dfrac{a}{u\sqrt{3}}$, A is at P, distance a due west of B, at point S.

$\mathbf{v}'_A$ is u in direction N. $60°$ E.;
$_A\mathbf{v}'_B = \mathbf{v}'_A - \mathbf{v}_B$.
From velocity triangle PQR,
$PQ = QR = u$, $\angle PQR = 60°$, therefore $PR = u$.
$_A\mathbf{v}'_B$ is u in direction S. $60°$ E.

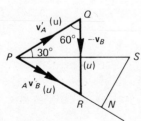

Shortest distance is $SN = a\sin 30° = \dfrac{a}{2}$.

Distance moved by A relative to $B = PN = \dfrac{a\sqrt{3}}{2}$.

Time taken $= \dfrac{a\sqrt{3}}{2u}$.

Distance moved by A is $\dfrac{a\sqrt{3}}{2}$.

Distance moved by A in the first part $= \dfrac{a}{u\sqrt{3}}\,(2u) = \dfrac{2a}{\sqrt{3}}$.

Hence total distance moved by A is $\dfrac{7a}{2\sqrt{3}}$.

2.5 In this question distances are measured in kilometres and speeds in km h^{-1}. i and j are perpendicular unit vectors.

A, B and C move in a plane with constant velocities and at time $t = 0$ the position vectors of A, B and C relative to an origin O are $i + 3j$, $9i + 9j$ and $6i + 13j$ respectively.

The velocity of C relative to A is $7i - 10j$ and of C relative to B is $9i - 12j$.

(a) Find the velocity vector of B relative to A. Show that A and B do not collide and find their shortest distance apart and the time when A and B are this distance apart.

(b) Show that B and C do collide and find the distance between A and C when this collision occurs. (SUJB)

● (a) $_C v_A = (7i - 10j)$, $_C v_B = (9i - 12j)$;

$_B v_A = {}_C v_A - {}_C v_B = -2i + 2j$.

The velocity vector of B relative to A is $-2i + 2j$.
Initial displacement vector $\overline{AB} = (9i + 9j) - (i + 3j)$
$\qquad\qquad\qquad\qquad = 8i + 6j$.
Since $_B v_A \neq k\overline{AB}$, where k is a constant, A and B will not collide.
Position vector of A after time $t = (i + 3j) + v_A t$.
Position vector of B after time $t = (9i + 9j) + v_B t$.
Displacement $\overline{AB} = (9i + 9j) + v_B t - (i + 3j) - v_A t$
$\qquad\qquad\qquad = 8i + 6j + t\,(-2i + 2j)$
$\qquad\qquad\qquad = (8 - 2t)\,i + (6 + 2t)\,j$.
$(AB)^2 = (64 - 32t + 4t^2) + (36 + 24t + 4t^2)$
$\qquad\quad = 100 - 8t + 8t^2$
$\qquad\quad = 8\,(t - 0.5)^2 + 98$.
Hence the least distance AB occurs when $t = 0.5$, when $AB = \sqrt{98} = 9.9$ km.
Alternative method
Least distance occurs when $_B v_A \cdot \overline{AB} = 0$ gives $t = 0.5$ and hence $AB_{\min} = 9.9$ km.

(b) Initial displacement $\overline{CB} = (9i + 9j) - (6i + 13j)$
$\qquad\qquad\qquad\qquad = 3i - 4j$.
$_C v_B = 9i - 12j = 3\,(3i - 4j)$.
Since $_C v_B = 3\overline{CB}$, B and C will collide after $\frac{1}{3}$ hours.
Displacement $\overline{CA}$ at time $t = (i + 3j) - (6i + 13j) + {}_A v_C t$
$\qquad\qquad\qquad\qquad = -5i - 10j - (7i - 10j)/3$
$\qquad\qquad\qquad\qquad = -(22i + 20j)/3$.
Distance AC is $\sqrt{\left[\left(\dfrac{22}{3}\right)^2 + \left(\dfrac{20}{3}\right)^2\right]} = 9.91$ km.

2.6 In this question the unit vectors i and j are due east and north respectively.

Two motorways intersect at a point O. Car A is on the first motorway travelling in the direction $8i + 15j$; car B is on the second motorway travelling in the direction $3i + 4j$.

At time $t = 0$ when car A is at the point O and has a constant speed of 85 km h^{-1} car B is still 2.5 km from O travelling at 100 km h^{-1}.

Write down, in terms of t where appropriate,

(a) the velocity of B relative to A,

(b) the position vector of each car relative to O,

(c) the position vector of B relative to A.

Given that visibility is 2.5 km, show that the cars are within sight of each other for just over 11 minutes.

● Unit vectors in the directions of the motorways are

$$\tfrac{8}{17}\mathbf{i} + \tfrac{15}{17}\mathbf{j} \quad \text{and} \quad \tfrac{3}{5}\mathbf{i} + \tfrac{4}{5}\mathbf{j}.$$

If B is still 2.5 km from O the initial position vector of B relative to O is

$$\tfrac{5}{2}\left(-\tfrac{3}{5}\mathbf{i} - \tfrac{4}{5}\mathbf{j}\right) = -1.5\mathbf{i} - 2\mathbf{j}.$$

$$\mathbf{v}_A = 85\left(\tfrac{8}{17}\mathbf{i} + \tfrac{15}{17}\mathbf{j}\right) = 40\mathbf{i} + 75\mathbf{j}, \qquad \mathbf{v}_B = 100\left(\tfrac{3}{5}\mathbf{i} + \tfrac{4}{5}\mathbf{j}\right) = 60\mathbf{i} + 80\mathbf{j}.$$

(a) Velocity of B relative to A is $\mathbf{v}_B - \mathbf{v}_A = 20\mathbf{i} + 5\mathbf{j}$.

(b) At time t the position vectors of A and B relative to O are:
$$\overline{OB} = -1.5\mathbf{i} - 2\mathbf{j} + t\,(60\mathbf{i} + 80\mathbf{j})$$
$$= (60t - 1.5)\mathbf{i} + (80t - 2)\mathbf{j} \text{ km}$$
$$\overline{OA} = 40t\mathbf{i} + 75t\mathbf{j} \text{ km}.$$

(c) The position vector of B relative to $A = \overline{OB} - \overline{OA} = (20t - 1.5)\mathbf{i} + (5t - 2)\mathbf{j}$.
$$(AB)^2 = (20t - 1.5)^2 + (5t - 2)^2$$
$$= 400t^2 - 60t + 2.25 + 25t^2 - 20t + 4$$
$$= 425t^2 - 80t + 6.25.$$
When $AB = 2.5$, $425t^2 - 80t + 6.25 = 6.25 \Rightarrow t(425t - 80) = 0$.
Hence $t = 0$ or $\dfrac{80}{425}$.

Therefore the cars are within sight for $\dfrac{80}{425}$ h or 11.3 min.

2.7   In this question $\mathbf{i}$ and $\mathbf{j}$ are vectors of magnitude 1 km in directions E. and N. respectively. Units of time and speed are hours and kilometres per hour.

A and B move in a horizontal plane, A with constant velocity $4\mathbf{i} + 4\mathbf{j}$ and B with constant acceleration $2\mathbf{i} + 2\mathbf{j}$. At time $t = 0$, A is at the point with position vector $\mathbf{i} + 4\mathbf{j}$ and B is at $4\mathbf{i} + \mathbf{j}$ moving with velocity $2\mathbf{j}$.

(a) Find the position vectors of A and B at time t and hence show that

$$\overline{AB} = (t^2 - 4t + 3)\mathbf{i} + (t^2 - 2t - 3)\mathbf{j}.$$

(b) Find the time when B will be due south of A and the distance AB at that moment.

(c) Show that A and B subsequently collide and give the time at which this happens.

(d) Find the magnitude and direction of the velocity of B just before the collision occurs.   (SUJB)

● (a) Let the position vectors of A and B be represented by $\mathbf{r}_A$ and $\mathbf{r}_B$ respectively.
$$\ddot{\mathbf{r}}_A = 0, \quad \mathbf{v}_A = \dot{\mathbf{r}}_A = 4\mathbf{i} + 4\mathbf{j}, \quad \mathbf{r}_A = (\mathbf{i} + 4\mathbf{j}) + (4\mathbf{i} + 4\mathbf{j})\,t.$$
∴ Position vector of A is $\mathbf{r}_A = (4t + 1)\mathbf{i} + (4t + 4)\mathbf{j}$,
$$\ddot{\mathbf{r}}_B = 2\mathbf{i} + 2\mathbf{j},$$
$$\mathbf{v}_B = \dot{\mathbf{r}}_B = 2\mathbf{j} + (2\mathbf{i} + 2\mathbf{j})\,t$$
$$= 2t\mathbf{i} + (2t + 2)\mathbf{j},$$
$$\mathbf{r}_B = (4\mathbf{i} + \mathbf{j}) + (t^2\mathbf{i} + (t^2 + 2t)\mathbf{j}).$$
Position vector of B is $\mathbf{r}_B = (t^2 + 4)\mathbf{i} + (t^2 + 2t + 1)\mathbf{j}$.
$$\overline{AB} = \mathbf{r}_B - \mathbf{r}_A = (t^2 - 4t + 3)\mathbf{i} + (t^2 - 2t - 3)\mathbf{j}.$$

(b) When B is due south of A, $t^2 - 4t + 3 = 0$, $t = 1$ or 3.
When $t = 1$, $\overline{AB} = -4\mathbf{j}$.
Therefore B is 4 km south of A after 1 h.
When $t = 3$, $\overline{AB} = 0$

(c) A and B collide after 3 h.

(d) When $t = 3$, $\mathbf{v}_B = 6\mathbf{i} + 8\mathbf{j}$; $|\mathbf{v}_B| = 10$.
Just before the collision B has velocity 10 km h^{-1}.
Direction of motion is N. arctan$\frac{6}{8}$ E., i.e. N. 36.9° E.

2.8   A cyclist A is pedalling at 3 m s^{-1} due east along a straight road. A second cyclist B is pedalling at 4 m s^{-1} due north along another straight road intersecting the first at a junction O. At a given instant when A is 80 m from O he observes B also approaching O but still 40 m away from O. A immediately accelerates at 0.1 m s^{-2} and B decelerates at q m s^{-2}. Find the velocity and position of B relative to A in terms of time t. Determine the value of q which causes them to arrive at O together.

- Let $\mathbf{n}$ and $\mathbf{e}$ be unit vectors due north and east, and $\mathbf{r}_A$ and $\mathbf{r}_B$ the position vectors of A and B.
  $\ddot{\mathbf{r}}_A = 0.1\mathbf{e}$, $\dot{\mathbf{r}}_A = (3 + 0.1t)\,\mathbf{e}$, $\mathbf{r}_A = (-80 + 3t + 0.05t^2)\,\mathbf{e}$
  $\ddot{\mathbf{r}}_B = -q\mathbf{n}$, $\dot{\mathbf{r}}_B = (4 - qt)\,\mathbf{n}$, $\mathbf{r}_B = (-40 + 4t - 0.5qt^2)\,\mathbf{n}$
  Velocity of B relative to $A = \dot{\mathbf{r}}_B - \dot{\mathbf{r}}_A = (4 - qt)\,\mathbf{n} - (3 + 0.1t)\,\mathbf{e}$.
  Position of B relative to $A = \mathbf{r}_B - \mathbf{r}_A$
  $$= (-40 + 4t - 0.5qt^2)\,\mathbf{n} - (-80 + 3t + 0.05t^2)\,\mathbf{e}.$$
  In order to arrive at O together, $80 - 3t - 0.05t^2 = 0$,
  i.e. $t^2 + 60t - 1600 = 0$
  i.e. $(t - 20)(t + 80) = 0$
  $\Rightarrow$ $t = 20$.
  Also $40 - 4t + 0.5qt^2 = 0$ when $t = 20$
  $\Rightarrow$ $40 - 80 + 200q = 0, q = 0.2$.

2.9   The captain of a ship A steaming due north at 12 knots sees a second ship B, distant 10 nautical miles due west, which appears to be moving in a direction 150° (S. 30° E.) at $12\sqrt{3}$ knots.
(a) Find the true velocity of ship B.
(b) Show that when they are closest to one another, the bearing of ship A from ship B is 060° (N. 60° E.).
(c) If A is a warship which has a firing distance of up to 9 nautical miles, prove that ship B could be liable to attack for just over 14 minutes.   (SUJB)

- From $\triangle XYW$, the relative velocity triangle,

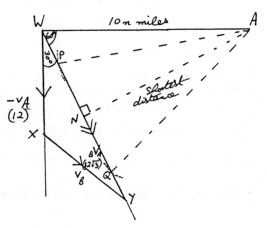

23

$$WX = |-\mathbf{v}_A| = 12, \qquad WY = |_B\mathbf{v}_A| = 12\sqrt{3}, \qquad W = 30°.$$

By the cosine rule,

$$XY \; (= |\mathbf{v}_B|) = \sqrt{[12^2 + (12\sqrt{3})^2 - 2(12)(12\sqrt{3})\cos 30°]}$$
$$= \sqrt{(12)^2} = 12.$$

Hence $\angle WYX = 30°$ and $\angle WXY = 120°$.

(a) The true velocity of ship B is 12 knots in direction S. 60° E.

(b) The ships are closest to one another when the displacement BA is perpendicular to the velocity of B relative to A, i.e. the distance AN on the sketch.

From $\triangle AWN$, $W = 60°$, $N = 90°$ so $A = 30°$.

Therefore the bearing of B from A is S. 60° W.

Shortest distance $= 10\sin 60° = 5\sqrt{3} = 8.66$ nautical miles.

(c) B is first in firing range when the displacement vector $\overline{BA}$ is $\overline{PA}$, where $PA = 9$.

From $\triangle APN$, $(PN)^2 = 9^2 - (5\sqrt{3})^2 = 6$.

Thus B is in firing range for $2\sqrt{6}$ nautical miles relative to A.

Since $|_B\mathbf{v}_A| = 12\sqrt{3}$ knots, time for which ship is in range is

$$\frac{(2\sqrt{6})\,60}{12\sqrt{3}} \text{ min} = 14.1 \text{ minutes.}$$

Construction:

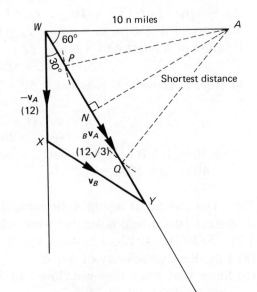

2.3 Exercises

2.1 The velocity of particle A is $2(\mathbf{i} - \mathbf{j})$ relative to particle B which moves with velocity $-4(\mathbf{i} + \mathbf{j})$. Find the speeds of both particles.

2.2 The vectors $\mathbf{i}$ and $\mathbf{j}$ are unit vectors in the directions east and north respectively. A man walks with constant speed 6 km h^{-1} due north and to him the wind appears to have a velocity $u_1[(\sqrt{3})\mathbf{i} - 3\mathbf{j}]$ km h^{-1}. Without changing speed the man alters course so that he is walking in the direction of the vector $-\dfrac{\sqrt{3}}{2}\mathbf{i} + \dfrac{1}{2}\mathbf{j}$ and the velocity of the wind now appears to him to be $u_2\,\mathbf{i}$ km h^{-1}. Find u_1 and u_2.

Find also the actual velocity of the wind. (L)

2.3 A man swims in a straight line due west from the end of a pier A to a buoy B, a distance of 2 km, and then back to the pier. The constant speed of the swimmer is $2\sqrt{3}$ km h^{-1} in still water, and he experiences a current of 2 km h^{-1} from 30° west of north. Find the directions in which the man must head on the outward and return journeys and the total time taken.

2.4 A model boat A, moving with constant velocity $2\mathbf{i} - 3\mathbf{j}$, passes through a point with position vector $0\mathbf{i} + 0\mathbf{j}$ at the same time as a second model boat B passes through a point with position vector $p\mathbf{i} + 2\mathbf{j}$. If B has a constant velocity $\mathbf{i} - 4\mathbf{j}$, find the position vectors of A and B at time t and deduce their relative position and the velocity of B relative to A. Find the value of p if the boats are on a collision course and the value of time t when they would collide.

If A changes velocity when $t = 1$ to $2\mathbf{i} + 3\mathbf{j}$, will the boats avoid collision?

2.5 Two particles P and Q are free to move in a horizontal plane; $\mathbf{i}$ and $\mathbf{j}$ denote perpendicular unit vectors in that plane and O is a fixed origin in the plane. The particles move with constant velocities $(9\mathbf{i} + 6\mathbf{j})$ m s^{-1} and $(5\mathbf{i} + 4\mathbf{j})$ m s^{-1} respectively.

Determine the speeds of the particles in metres per second, correct to one decimal place.

At time $t = 4$ s the particles P and Q have position vectors, referred to O, of $(96\mathbf{i} + 44\mathbf{j})$ m and $(100\mathbf{i} + 96\mathbf{j})$ m, respectively.

Find:
(a) the position vectors of P and Q at time $t = 0$ s,
(b) the vector PQ at time t,
(c) the time at which P and Q are nearest to each other and the length of PQ, in metres to one decimal place, at this instant. (AEB)

2.6 An aircraft A is travelling due west at 240 mile h^{-1}. At noon a second aircraft B, at the same altitude, is 40 miles away from A in a southwesterly direction. One hour later B is again 40 miles away but now in a north-westerly direction. Calculate the position vector of B relative to A at time t. Deduce the minimum distance between A and B and show that this is when B is due west of A. Find the velocity of B.

2.7 In this question distances are measured in metres and time in seconds.

At time $t = 0$ two particles P and Q are set in motion in the x–y plane. Initially P is at A $(1, 0)$ and Q is at B $(0, 8)$. The particle P moves with a constant speed of 5 m s^{-1} parallel to the line $3y = 4x$ and Q moves with a constant speed of 4 m s^{-1} parallel to the line $y = -\lambda x$, the sense of motion of both P and Q being that in which x is increasing. Given that $\mathbf{i}$ and $\mathbf{j}$ are the unit vectors in the directions of x increasing and y increasing, respectively, show that the unit vectors in the directions of motion of P and Q are $\frac{3}{5}\mathbf{i} + \frac{4}{5}\mathbf{j}$ and $\frac{1}{\sqrt{(1+\lambda^2)}}\mathbf{i} - \frac{1}{\sqrt{(1+\lambda^2)}}\mathbf{j}$ respectively.

Determine, in the form $a\mathbf{i} + b\mathbf{j}$,
(a) the velocities of P and Q,
(b) the vectors $\overline{AP}$, $\overline{BQ}$ and $\overline{PQ}$ at time t.

Show that, if P and Q meet, λ must satisfy the equation

$$7(1 + \lambda^2)^{1/2} = 8 - \lambda.$$

Verify that $\lambda = -\frac{3}{4}$ is a solution of this equation and for this value of λ find the time when P and Q meet.

(AEB 1982)

2.4 Brief Solutions to Exercises

2.1 $_A\mathbf{v}_B = \mathbf{v}_A - \mathbf{v}_B$ so $\mathbf{v}_A = {}_A\mathbf{v}_B + \mathbf{v}_B$
$$= 2(\mathbf{i} - \mathbf{j}) + -4(\mathbf{i} + \mathbf{j}) = -2\mathbf{i} - 6\mathbf{j}.$$

Speed of $A = \sqrt{(4 + 36)} = 2\sqrt{10} = 6.32$,   speed of $B = 4\sqrt{2} = 5.66$.
Alternatively, by construction:

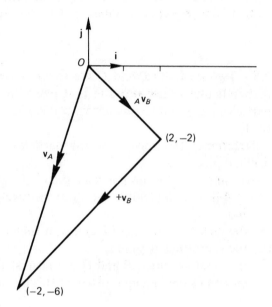

2.2 $_W\mathbf{v}_M = \mathbf{v}_W - \mathbf{v}_M$ so $\mathbf{v}_W = {}_W\mathbf{v}_M + \mathbf{v}_M$
$$= u_1 \, [(\sqrt{3})\mathbf{i} - 3\mathbf{j}] + 6\mathbf{j} \qquad \text{(1st walk)}$$
$$= u_2 \mathbf{i} + 6\left(\frac{-\sqrt{3}}{2}\mathbf{i} + \tfrac{1}{2}\mathbf{j}\right). \qquad \text{(2nd walk)}$$

Equating $\mathbf{i}$ components, $\sqrt{3}u_1 = u_2 - 3\sqrt{3}$.   (1)
Equating $\mathbf{j}$ components, $6 - 3u_1 = 3$.   (2)
Solving, $u_1 = 1$ and $u_2 = 4\sqrt{3}$. Velocity of wind $= (\sqrt{3})\mathbf{i} + 3\mathbf{j}$.

2.3   *Outward journey*

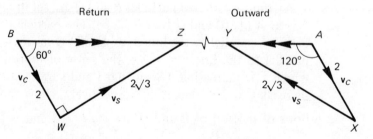

$\triangle AXY$, $AX =$ speed of current $= 2$,
$XY =$ speed of swimmer in still water $= 2\sqrt{3}$, $AY =$ true speed, $A = 120°$.
By the sine rule, $Y = 30°$, $X = 30°$,   so $AY = 2$.
Swimmer travels at 2 km h^{-1} and heads N. 60° W.

Return journey
$\triangle BWZ$, $BW = 2$, $WZ = 2\sqrt{3}$, BZ = true speed, $B = 60°$.
By the sine rule, $Z = 30°$,   so $W = 90°$ and $BZ = 4$.
Swimmer travels at 4 km h^{-1}, heading N. 60° E. Total time 1 h 30 min.

2.4 $v_A = 2i - 3j$, $\overline{OA} = 2ti - 3tj$.
   $v_B = i - 4j$, $\overline{OB} = pi + 2j + ti - 4tj$
                $= (p + t)i + (2 - 4t)j$.
$\overline{AB} = \overline{OB} - \overline{OA} = (p - t)i + (2 - t)j$
       $_Bv_A = -i - j$.
Collide if/when $AB = 0$, then $t = 2$ so $p = 2$.
When $t = 1$, $\overline{OA} = 2i - 3j$ and $\overline{OB} = 3i - 2j$.
After further time t_1, $\overline{OA} = 2i - 3j + t_1 (2i + 3j)$,
          $\overline{OB} = 3i - 2j + t_1 (i - 4j)$,
          $\overline{AB} = (1 - t_1)i + (1 - 7t_1)j$.
This is never zero, so collision is avoided.

2.5 $v_P = (9i + 6j)$,   speed $= \sqrt{(9^2 + 6^2)} = 10.8$ m s^{-1},
$r_P = (9t + c_1)i + (6t + c_2)j$.
$v_Q = (5i + 4j)$,   speed $= \sqrt{(5^2 + 4^2)} = 6.4$ m s^{-1}, $r_Q = (5t + c_3)i + (4t + c_4)j$.
Conditions when $t = 4$ give $c_1 = 60$, $c_2 = 20$, $c_3 = 80$, $c_4 = 80$.
(a) When $t = 0$, $r_P = 60i + 20j$, $r_Q = 80i + 80j$.
(b) $\overline{PQ} = r_Q - r_P = (20 - 4t)i + (60 - 2t)j$.
(c) $PQ = \sqrt{(400 - 160t + 16t^2 + 3600 - 240t + 4t^2)} = \sqrt{[20(t - 10)^2 + 2000]}$.
   Minimum distance occurs when $t = 10$, then $\overline{PQ} = 20\sqrt{5}$ m $= 44.7$ m.
   Alternative method:
   Use $\overline{PQ} \cdot {}_Pv_Q = 0$ at minimum displacement.

2.6   At $t = 0$, $\overline{AB} = 20(\sqrt{2})w + 20(\sqrt{2})s$.
After 1 h, $\overline{AB} = 20(\sqrt{2})w - 20(\sqrt{2})s$.

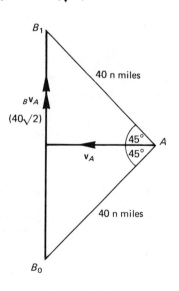

$_Bv_A = -40(\sqrt{2})s$ $\Rightarrow$ $\overline{AB} = 20(\sqrt{2})w + 20(\sqrt{2})s - 40(\sqrt{2})t s$
                $= 20\sqrt{2}[w + (1 - 2t)s]$.
Minimum $|\overline{AB}|$, when $t = 0.5$ is $\overline{AB} = 20(\sqrt{2})w$,
magnitude $20\sqrt{2}$ miles and direction west.

$\mathbf{v}_B = {}_B\mathbf{v}_A + \mathbf{v}_A = -40\,(\sqrt{2})\,\mathbf{s} + 240\,\mathbf{w}$, speed 246.6 mile h^{-1},

direction N. 76.7° W.

Construction:

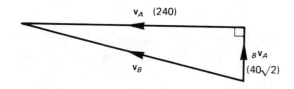

2.7 A unit vector at $\theta°$ to the positive x-axis is $\cos\theta\,\mathbf{i} + \sin\theta\,\mathbf{j}$. From the diagram, $\tan\theta = \frac{4}{3}$, unit vector $= \frac{3}{5}\mathbf{i} + \frac{4}{5}\mathbf{j} = \mathbf{l}_1$.

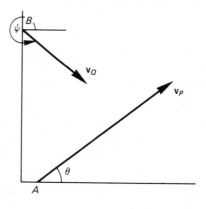

$\tan\psi = -\lambda$, unit vector $= \dfrac{1}{\sqrt{(1+\lambda^2)}}\mathbf{i} - \dfrac{\lambda}{\sqrt{(1+\lambda^2)}}\mathbf{j} = \mathbf{l}_2$.

(a) $\mathbf{v}_P = 5\mathbf{l}_1 = 3\mathbf{i} + 4\mathbf{j}$, $\mathbf{v}_Q = 4\mathbf{l}_2 = \dfrac{4}{\sqrt{(1+\lambda^2)}}\mathbf{i} - \dfrac{4\lambda}{\sqrt{(1+\lambda^2)}}\mathbf{j}$.

(b) $\overline{AP} = 3t\mathbf{i} + 4t\mathbf{j}$, $\overline{BQ} = \dfrac{4t}{\sqrt{(1+\lambda^2)}}\mathbf{i} - \dfrac{4t\lambda}{\sqrt{(1+\lambda^2)}}\mathbf{j}$.

$\overline{PQ} = \overline{PA} + \overline{AO} + \overline{OB} + \overline{BQ}$

$\qquad = \left(\dfrac{4t}{\sqrt{(1+\lambda^2)}} - 1 - 3t\right)\mathbf{i} + \left(8 - 4t - \dfrac{4\lambda t}{\sqrt{(1+\lambda^2)}}\right)\mathbf{j}$.

To meet, $\overline{PQ} = 0$ for some value of t. Equate components to zero.

$\dfrac{4t}{\sqrt{(1+\lambda^2)}} = 1 + 3t$ $\quad$ (1), $\qquad \dfrac{\lambda t}{\sqrt{(1+\lambda^2)}} = 2 - t$. $\quad$ (2)

$2(1) - (2)$: $\qquad 8t - \lambda t = 7t\sqrt{(1+\lambda^2)}$. Hence $7\sqrt{(1+\lambda^2)} = 8 - \lambda$.

If $\lambda = -\frac{3}{4}$, l.h.s. $= \frac{35}{4}$, r.h.s. $= \frac{35}{4}$. Hence $\lambda = -\frac{3}{4}$ satisfies equation.

Substitute $\lambda = -\frac{3}{4}$ into (1) or (2) to get $t = 5$.

3 Systems of Coplanar Forces. Frameworks

Forces as vectors. Composition and resolution of coplanar forces. Resultant of coplanar forces. Triangle and polygon of forces. Moment of a force about a point. (Not vector products.)

Couples. Reduction of a coplanar system of forces to a force at a point and a couple. Problems may require clear force diagrams. Graphical solutions may be permitted where appropriate.

3.1 Fact Sheet

(a) Definitions

(i) The component of a force $\mathbf{F}$, having magnitude F, in a direction making θ with $\mathbf{F}$ is $F \cos \theta$. Forces are usually resolved into two perpendicular components F_x and F_y.

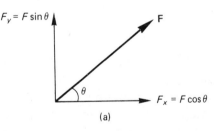

(a)

(ii) The moment M of a force $\mathbf{F}$ about a point P is Fd where d is the perpendicular distance of P from the line of action of $\mathbf{F}$.

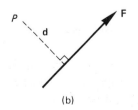

(b)

(iii) A pair of forces equal in magnitude and opposite in direction form a couple. The moment or magnitude of a couple is the product of the magnitude of either force and the perpendicular distance between them.

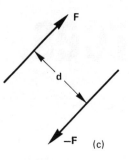

(b) **Resultant of Forces**

(i) If forces P_1, P_2, . . ., are represented by the sides $AB, BC, . . ., MN$ of a polygon, then the last side AN represents the resultant $\mathbf{R}$ in magnitude and direction. If there are only two forces this becomes a triangle of forces; if more a polygon of forces.

(ii) If the components of the forces have been found then

$$R^2 = (\Sigma F_x)^2 + (\Sigma F_y)^2 \qquad \text{and} \qquad \tan \theta = \frac{\Sigma F_y}{\Sigma F_x}.$$

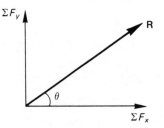

(iii) If the forces all act at one point then the resultant force acts at that point.

(iv) If the forces act in a plane, not all at one point, then the magnitude and direction of the resultant is found as above and a point on the line of action is found by taking moments about any point in the plane.

(v) If $R = 0$ and $M \neq 0$, then the system of forces reduces to a couple of magnitude M.

(vi) If $R = 0$ and $M = 0$ then the system of forces is in equilibrium.

(c) **Equilibrium of Three Forces**

If three forces are in equilibrium then they must be parallel, or pass through one point. If they pass through one point then they can be represented by the sides of a triangle taken in order.

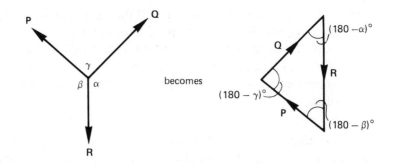

Lami's theorem, $\dfrac{P}{\sin\alpha} = \dfrac{Q}{\sin\beta} = \dfrac{R}{\sin\gamma}$, is equivalent to the sine rule.

(d) Light Frameworks

(i) Each joint is in equilibrium.

(ii) The rods are usually light — neglect weight.

(iii) Each rod exerts equal and opposite forces at its ends on the joints. These are equal and opposite to those exerted by the joints on the rods.

Arrows on force diagrams usually indicate the forces on the joints:

$\longleftarrow\!\!\!\!\!\longrightarrow$ implies that the rod is under compression;

$\longrightarrow\!\!\!\!\!\longleftarrow$ implies that the rod is in tension.

(e) Methods of Solution of Light Frameworks

(i) If possible find external forces first by taking moments and/or resolving.

(ii) Then either

(1) resolve forces in two perpendicular directions at each joint and solve the resulting equations, or

(2) use Bow's notation and draw the force polygons for successive joints (forces are parallel to the rods). Since the forces are in equilibrium each polygon must be closed.

Hint: Start at a joint with no more than two unknown forces.

3.2 Worked Examples

3.1 Show that an arbitrary system of forces in a plane is, in general, equivalent to a single force and a couple.

If the single force is non-zero, show that there is a line in the plane about which the system of forces has no moment.

● Let all the forces have components X_i, Y_i and let the sum of the moments about any point A be M_A.

(a) If ΣX_i and/or $\Sigma Y_i \neq 0$ the system is equivalent to a single force of magnitude $R = \sqrt{[(\Sigma X_i)^2 + (\Sigma Y_i)^2]}$. This can be expressed as a force $\mathbf{R}$ and a couple by introducing forces $\mathbf{R}$ and $-\mathbf{R}$ at any point.

(b) If ΣX_i and $\Sigma Y_i = 0$ and $M_A \neq 0$ then the system is equivalent to a force of zero magnitude and a couple of moment M_A.

    If $R \neq 0$ then the line of action of force **R** can be found by $M_A = Rd$ where d is the perpendicular distance of A from the line of action of **R**. The system of forces then has no moment about any point on the line of action of **R**.

3.2   A uniform hexagonal lamina $ABCDEF$ has weight W. The lamina is smoothly pivoted at B. It is kept in equilibrium in a vertical plane with AB horizontal and the vertices C, D, E and F above AB by a force of magnitude P acting at D in the direction of AD. Find P and the magnitude and direction of the force exerted on the lamina by the pivot.

    The force at D is reversed in direction while retaining its magnitude. Find the weight which must be attached at C to keep the lamina in this position of equilibrium.   (L)

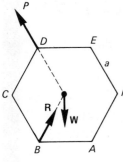

- Let the side of the hexagon be a.
  Taking moments about B,

$$W\frac{a}{2} = P\frac{a\sqrt{3}}{2} \qquad \text{so} \qquad P = \frac{W}{\sqrt{3}}.$$

Force R at the pivot must pass through the centre of the hexagon, since the other forces have no moment about the centre.
    Resolving perpendicular to DA:

$$R\cos 30 = W\cos 60 \qquad \text{so} \qquad R = \frac{W}{\sqrt{3}}.$$

Let weight w be attached at C when P is reversed. Taking moments about B,

$$w\frac{a}{2} = W\frac{a}{2} + \frac{W}{\sqrt{3}}\frac{a\sqrt{3}}{2}, \qquad \text{so} \qquad w = 2W.$$

A weight of $2W$ must be attached at C for equilibrium.

3.3   A rectangle $ABCD$ has sides $AB = 3a$, $BC = 4a$. Forces of magnitude $7W$, $6W$, $10W$, $13W$ and $15W$ act along the lines BA, BC, DC, DA and AC respectively in the directions indicated by the order of the letters. Find the resultant of this system in magnitude and direction and the distance from A at which its line of action cuts AD.

    An extra force P is now added at D so that the system of forces reduces to a couple. Find the value of P and the magnitude of the resulting couple.

- From given dimensions, $\tan\alpha = \frac{3}{4}$, $\sin\alpha = \frac{3}{5}$, $\cos\alpha = \frac{4}{5}$.

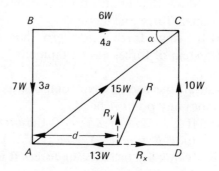

Resolving forces, parallel to AD, $R_x = 6W + 15W \cos \alpha - 13W = 5W$,
parallel to AB, $R_y = 10W + 15W \sin \alpha - 7W = 12W$.
Hence the resultant force is given by

$$R = \sqrt{[(5W)^2 + (12W)^2} = 13W,$$

making an angle $\tan^{-1} \frac{5}{12}$ with AB, i.e. $22.6°$ with AB.
Let the line of action of R cut AD at a distance d from A.

Taking moments about A, $(10W)(4a) - (6W)(3a) = R_y d = (12W)(d)$,
$(22W)(a) = (12W)(d)$, so $d = \frac{11}{6}a$.

The line of action of the resultant force cuts AD at a distance $\frac{11}{6}a$ from A.

If the resultant of P and R is a couple, then P is an equal and opposite force to R, i.e. P has a magnitude of $13W$.

The magnitude of the resulting couple is its moment about D.
Magnitude $= R_y (4a - d) = 12W (\frac{13}{6}) = 26Wa$.

Hint: It is usually easier to use the components of R when taking moments, rather than calculating the perpendicular distance from A to the line of action of R.

In this case with the line of action of R cutting AD, R has components
$R_x = 5W$ along AD (with no moment about A),
and $R_y = 12W$ perpendicular to AD (with moment $12W(d)$ about A).

3.4 Two forces **P** and **Q** are such that the resultant of **P** and **Q** is perpendicular to **P** and the resultant of **2P** and **Q** is perpendicular to **Q**. Prove that the resultant of **P** and **Q** has magnitude P and the resultant of **2P** and **Q** has magnitude Q.

Find the ratio $Q : P$ and the angle between **P** and **Q**. (OLE)

● Resultant $\mathbf{R_1}$ of **P** and **Q** is **P** + **Q**.
Since $\mathbf{R_1}$ is perpendicular to **P** then $\mathbf{R_1} \cdot \mathbf{P} = 0$,

i.e. $P^2 + \mathbf{P} \cdot \mathbf{Q} = 0$. (1)

Similarly, $\mathbf{R_2} = 2\mathbf{P} + \mathbf{Q}$ and $\mathbf{R_2} \cdot \mathbf{Q} = 0$,

i.e. $2\mathbf{P} \cdot \mathbf{Q} + Q^2 = 0$. (2)

$R_1^2 = (\mathbf{P} + \mathbf{Q}) \cdot (\mathbf{P} + \mathbf{Q}) = P^2 + Q^2 + 2\mathbf{P} \cdot \mathbf{Q}$.

From (2), $2\mathbf{P} \cdot \mathbf{Q} + Q^2 = 0 \Rightarrow R_1^2 = P^2$.

So the magnitude of $\mathbf{R_1}$ is P.

$R_2^2 = (2\mathbf{P} + \mathbf{Q}) \cdot (2\mathbf{P} + \mathbf{Q}) = 4P^2 + Q^2 + 4\mathbf{P} \cdot \mathbf{Q}$.

From (1), $4P^2 + 4\mathbf{P} \cdot \mathbf{Q} = 0 \Rightarrow R_2^2 = Q^2$.

So the magnitude of $\mathbf{R_2}$ is Q.

From (1), $\mathbf{P} \cdot \mathbf{Q} = -P^2$.

Substitute into (2): $-2P^2 + Q^2 = 0$.

Therefore $\dfrac{Q^2}{P^2} = 2$ and $Q : P = \sqrt{2} : 1$.

(Magnitudes are essentially positive.)

By scalar products, $\mathbf{P} \cdot \mathbf{Q} = PQ \cos \theta$, where θ is the angle between **P** and **Q**,

$\Rightarrow -P^2 = (\sqrt{2})P^2 \cos \theta$, $\cos \theta = -\dfrac{1}{\sqrt{2}}$, $\theta = \frac{3}{4}\pi$.

Therefore the angle between **P** and **Q** is $\frac{3}{4}\pi$.

3.5 Forces act on a lamina in the shape of a trapezium $ABCD$ with AB parallel to DC, and $\overline{DC} = 2\overline{AB}$. Determine in each case the additional force necessary to produce equilibrium and determine one point on the lamina through which the line of action of the force must pass.
(a) Forces represented by $\overline{AB}$, $\overline{AD}$, $\overline{DC}$, and $\overline{BC}$ and acting along the corresponding lines.
(b) Forces represented by $\overline{AC}$, $\overline{AD}$ and $\overline{CD}$ and acting along the corresponding lines.
(c) Forces represented by $\overline{BA}$, $2\overline{DB}$ and $\overline{AD}$ and acting along the corresponding lines.

● (a) $\overline{AB} + \overline{BC} = \overline{AC}$,

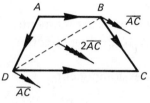

i.e., the resultant of $\overline{AB}$ and $\overline{BC}$ is a force through B, equal in magnitude and direction to $\overline{AC}$.
$\overline{AD} + \overline{DC} = \overline{AC}$,
i.e., the resultant of $\overline{AD}$ and $\overline{DC}$ is a force through D, equal in magnitude and direction to $\overline{AC}$.
Therefore the resultant of forces $\overline{AB}$, $\overline{AD}$, $\overline{DC}$ and $\overline{BC}$ is of magnitude $2\overline{AC}$, passing through the mid-point of BD.

The additional force needed for equilibrium is $2\overline{CA}$, passing through the mid-point of BD.

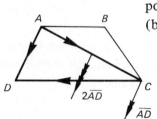

(b) $\overline{AC} + \overline{CD} = \overline{AD}$,

i.e., the resultant of $\overline{AC}$ and $\overline{CD}$ is a force through C, equal in magnitude and direction to $\overline{AD}$.

Therefore the resultant of $\overline{AC}$, $\overline{CD}$ and $\overline{AD}$ is a force of magnitude $2\overline{AD}$, passing through the mid-points of DC and AC.

The additional force needed for equilibrium is $2\overline{DA}$, passing through the mid-point of DC.

(c) $\overline{BA} + \overline{AD} = \overline{BD}$,

i.e., the resultant of $\overline{BA}$ and $\overline{AD}$ is a force through A, equal in magnitude and direction to $\overline{BD}$.

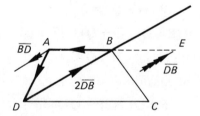

Therefore $\overline{BA} + \overline{AD} + 2\overline{DB} = \overline{BD} + 2\overline{DB} = \overline{DB}$.

By taking moments about A the resultant force cuts AB produced at E where $AE = 2AB$.

The additional force needed for equilibrium is $\overline{BD}$, passing through E.

3.6 Forces of magnitude 2 N, 4 N and 4 N act in the sense indicated by the letters along the sides AB, BC, CA respectively of an equilateral triangle ABC of side 2 m. Find the magnitude and direction of their resultant and the point where its line of action cuts AC produced.

This system of forces is to be reduced to equilibrium by the addition of a couple in the plane ABC and a force which acts through A. Find the magnitude and direction of the force and the magnitude and sense of the couple. (L)

● The resultant of 4 N along BC and 4 N along CA is 4 N, parallel to BA, passing through C.

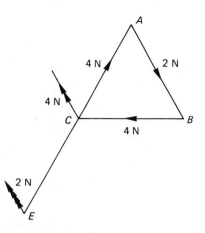

The resultant of this force at C and 2 N along AB is 2 N parallel to BA.
Let the line of action cut AC produced at E.
Taking moments about C, $(2)(\sqrt{3}) = (2)(CE \sin 60°)$, so $CE = 2$ m.
Reduce to equilibrium by introducing a couple of two equal and opposite forces of 2 N at E and A and a single force of 2 N at A along AB.

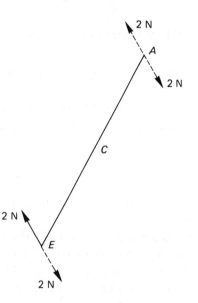

Therefore the force has magnitude 2 N along AB and the couple has magnitude $(2)(4 \sin 60°)$ N m, i.e. $4\sqrt{3}$ N m. Sense $A \rightarrow C \rightarrow B \rightarrow A$.

3.7 A light rod AB, of length $2l$, is hinged to a vertical wall at A and is supported in a horizontal position, perpendicular to the wall, by an inextensible string joining B to a point C vertically above A so that $\angle ABC = 60°$.
Loads of weight W are hung from the mid-point of AB and from B.
Find the tension in the string and the magnitude and direction of the reaction at the hinge.

● Let the tension be T and the horizontal and vertical components of the reaction at A be R and S respectively.

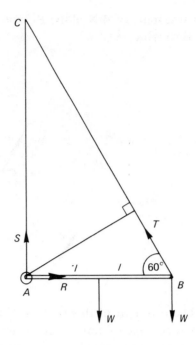

Taking moments about A, $(T)(2l \sin 60°) = Wl + W(2l)$.

So $T = \dfrac{3}{\sqrt{3}} W = \sqrt{3}W$.

Resolving horizontally, $R = T \cos 60° = \dfrac{\sqrt{3}}{2} W$.

Resolving vertically, $S + T \sin 60° = 2W \Rightarrow S = 2W - \sqrt{3}W\left(\dfrac{\sqrt{3}}{2}\right) = \dfrac{W}{2}$.

Thus the reaction at the hinge is $\sqrt{(R^2 + S^2)} = W$ at an angle θ to the rod where $\tan \theta = \dfrac{S}{R} = \dfrac{1}{\sqrt{3}}$, i.e. at $30°$ to the rod.

3.8 In this question all answers should be given correct to three significant figures.

ABC is a triangle with $A = 30°$ and $C = 40°$.

(a) Find the magnitude of the forces acting at A in directions $\overline{AB}$ and $\overline{BC}$ which are equivalent to a force of 20 N acting along AC.

(b) A force of 30 N acts along AC and another force **P** of magnitude 30 N acts at A. What is the direction of **P** if the resultant acts in the direction $\overline{BC}$ and what is the magnitude of the resultant?

(c) A force of 40 N acts along AC and a force **Q** along BA. What is the magnitude of **Q** if the resultant acts in a direction perpendicular to BC? (SUJB)

● (a) If the resultant of the forces acts along AC then forces must be proportional to the sides of the triangle.

If the forces are $\mathbf{F}_1$ and $\mathbf{F}_2$ along AB and BC respectively,

then $\dfrac{F_1}{\sin 40°} = \dfrac{F_2}{\sin 30°} = \dfrac{20}{\sin 110°}$,

giving $F_1 = 13.7$ N (to 3 s.f.),

and $F_2 = 10.6$ N (to 3 s.f.).

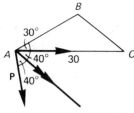

(b) Resultant of two equal forces bisects the angle between the forces, therefore P acts at $80°$ to AC, on the opposite side of AC from B.

Resolving the forces along the bisector of the angle,

$$R = 2P \cos 40° = 46.0 \text{ N (to 3 s.f.)}.$$

(c) If the resultant acts perpendicular to BC, resolving parallel to BC gives:

$$Q \cos 70° = 40 \cos 40° \quad \Rightarrow \quad Q = 89.6 \text{ N (to 3 s.f.)}.$$

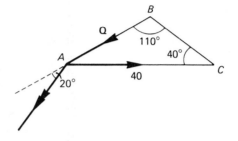

3.9 $\mathbf{P}$ and $\mathbf{Q}$ are two perpendicular forces and the magnitude of $\mathbf{Q}$ is twice that of $\mathbf{P}$. By **vector methods**, prove that the magnitude of the resultant of $\mathbf{P}$ and $\mathbf{Q}$ is equal to the magnitude of the resultant of $\mathbf{P}$ and $-\mathbf{Q}$ and find the angle between these two resultants.

Also prove that $(\mathbf{P} + 2\mathbf{Q})^2 + (\mathbf{P} - 3\mathbf{Q})^2 = 14P^2 + 10Q^2$.　　　　　(OLE)

● Let the resultant of $\mathbf{P}$ and $\mathbf{Q}$ be $\mathbf{R}_1$, and that of $\mathbf{P}$ and $-\mathbf{Q}$ be $\mathbf{R}_2$.

$$\mathbf{R}_1 = \mathbf{P} + \mathbf{Q}, \qquad \mathbf{R}_2 = \mathbf{P} - \mathbf{Q}, \qquad R_1^2 = \mathbf{R}_1 . \mathbf{R}_1 = (\mathbf{P} + \mathbf{Q}) . (\mathbf{P} + \mathbf{Q})$$
$$= P^2 + Q^2 + 2\mathbf{P} . \mathbf{Q}$$

where $Q = 2P$ (given).

But $\mathbf{P}$ and $\mathbf{Q}$ are perpendicular, so $\mathbf{P} . \mathbf{Q} = 0$.

Therefore $R_1^2 = P_2 + 4P^2 = 5P^2$ and $R_1 = \sqrt{5}P$.

Similarly,

$$R_2^2 = \mathbf{R}_2 . \mathbf{R}_2 = P^2 + Q^2 - 2\mathbf{P} . \mathbf{Q} = P^2 + 4P^2, \quad \text{so} \quad R_2 = \sqrt{5}P = R_1 \text{ as required}.$$

By scalar products, $\mathbf{R}_1 . \mathbf{R}_2 = P^2 - Q^2 = -3P^2$.

But $\mathbf{R}_1 . \mathbf{R}_2 = R_1 R_2 \cos\theta$ where θ is the angle between $\mathbf{R}_1$ and $\mathbf{R}_2$.

Thus $\cos\theta = \dfrac{-3P^2}{5P^2} = -\dfrac{3}{5}$, giving an angle between the resultants of $126.9°$.

$$(\mathbf{P} + 2\mathbf{Q})^2 + (\mathbf{P} - 3\mathbf{Q})^2 = P^2 + 4\mathbf{P} . \mathbf{Q} + 4Q^2 + P^2 - 6\mathbf{P} . \mathbf{Q} + 9Q^2.$$

However, P and Q are perpendicular, so $\mathbf{P} . \mathbf{Q} = 0$ and

$$(\mathbf{P} + 2\mathbf{Q})^2 + (\mathbf{P} - 3\mathbf{Q})^2 = 2P^2 + 13Q^2 = 14P^2 + 10Q^2, \qquad \text{since} \qquad Q = 2P.$$

3.10 The figure shows a framework which comprises six light rods $AB, BC, BE,$ BD, ED and DC freely jointed. The points A and E are fixed so that EA is vertical. Each of the inclined rods make an angle of $60°$ with the horizontal. C carries a load of 1 kN and there is a horizontal force of 2 kN at B. Explain why the force in the rod AB at A is in the direction AB. Find, graphically or otherwise

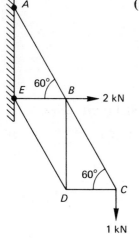

(a) the reactions at the joints A and E in magnitude and direction,

(b) the stresses in the rods EB, BD and ED, distinguishing between compression and tension.

(SUJB)

● Let the reactions at A and E be R and P respectively and all internal forces as shown in the diagram. All forces in kN.

Taking moments about E, the only external forces which have a turning moment about E are 1 kN at C and the reaction R at A.

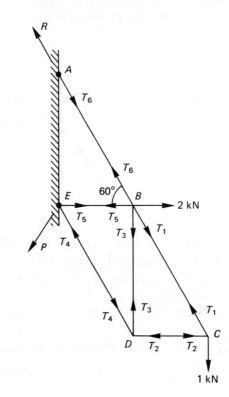

1 kN has a clockwise moment about E and therefore R must have an anti-clockwise moment of equal magnitude. Thus R acts in the direction BA. For equilibrium the force in the rod must act in the direction AB.

Method 1: Calculation

Forces at C: $T_1 \cos 30° = 1$ $\Rightarrow$ $T_1 = \dfrac{2}{\sqrt{3}} = 1.15$ kN,

$$T_2 = T_1 \sin 30° \quad \Rightarrow \quad T_2 = \dfrac{1}{\sqrt{3}} = 0.58 \text{ kN}.$$

Forces at D: $T_4 \cos 60° = T_2$ $\Rightarrow$ $T_4 = \dfrac{2}{\sqrt{3}} = 1.15$ kN,

$$T_4 \cos 30° = T_3 \quad \Rightarrow \quad T_3 = 1 \text{ kN}.$$

Forces at B: $T_6 \cos 30° = T_1 \cos 30° + T_3$ $\Rightarrow$ $T_6 = \dfrac{4}{\sqrt{3}} = 2.31$ kN,

$$T_5 + T_6 \cos 60° = 2 + T_1 \cos 60° \quad \Rightarrow \quad T_5 = 2 - \dfrac{1}{\sqrt{3}} = 1.42 \text{ kN}.$$

Forces at E: the reaction P at E has components

$$T_5 - T_4 \cos 60° \text{ horizontally} \quad \text{i.e.} \quad 2 - \dfrac{2}{\sqrt{3}} = 0.85 \text{ kN}$$

and $T_4 \cos 30°$ vertically   i.e.   1 kN.

Therefore the reaction at E is 1.31 kN at 49.8° below the horizontal.
Reaction at A is 2.31 kN in direction BA.
Stress in EB is 1.42 kN (in tension).
Stress in BD is 1 kN (in tension).
Stress in ED is 1.15 kN (in compression).
Method 2: Bow's notation

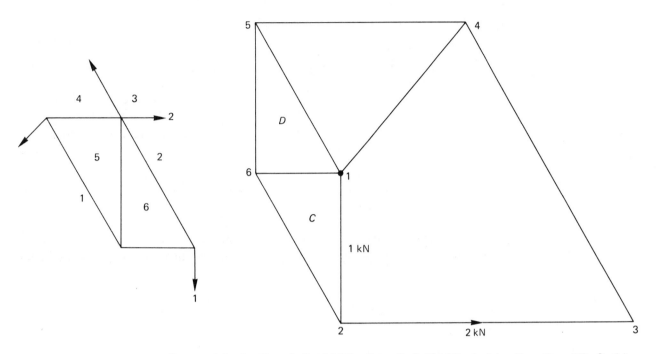

Start with the line 1–2 (1 kN), then 2–3 (2 kN), 1–6 in direction CD, 2–6 in direction CB, 6–5 in direction DB, 1–5 in direction DE, 5–4 in direction EB, 3–4 in direction BA. 4–1 gives P.

By measurement, reaction at A is given by the line 3–4 (2.33 kN) etc.

3.3 Exercises

3.1 $OABC$ is a square of side 1 m. Forces of magnitudes 4, 3, 2 and 5 newtons act along the sides OC, CB, AB and OA respectively in the direction indicated by the direction of the letters.
(a) Find the magnitude of the single force to which the system is equivalent and the equation of its line of action referred to OA and OC as axes of x and y respectively.
(b) Find the magnitude and direction of the force which must be applied to the system at O in order that it should reduce to a couple and give the moment and sense of the couple.
Instead of the force applied in (b) a couple is applied to the system so that the resultant then passes through O. Find the moment and sense of the couple and the magnitude and direction of the resultant.   (SUJB)

3.2 $ABCD$ is an isosceles trapezium $AD = DC = CB = a$ and $AB = 2a$. Five forces, measured in newtons and of magnitudes 1, 3, 5, 6 and $2\sqrt{3}$, act along AD, DC, CB, BA and AC respectively, the direction of each force being shown by the order of the letters.
The resultant force has magnitude R and its line of action cuts AB produced at X. Find the magnitude and direction of R and the distance AX.

3.3 (multiple choice)   A uniform rod PQ, of mass 6 kg, is smoothly hinged to a vertical wall at P. The rod makes an angle of 60° with the vertical. It is kept in equilibrium by a horizontal force of magnitude T newtons acting at Q. T is approximately equal to:

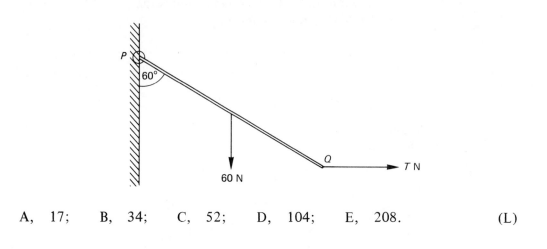

A, 17;   B, 34;   C, 52;   D, 104;   E, 208.   (L)

3.4   The diagram shows a uniform rod AB of weight W resting at an angle θ to the horizontal with its lower end in contact with a smooth vertical wall which is

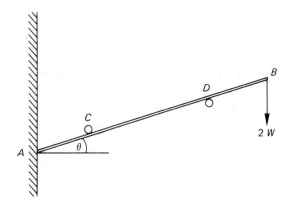

perpendicular to the vertical plane containing the rod. The rod passes under a peg C and over a peg D, both pegs being fixed, smooth, horizontal and parallel to the wall. A weight $2W$ is suspended from B and the rod rests in equilibrium. Given that $AC = DB = a$ and $CD = 2a$, find the reactions at C and D and show that $\cos^2 \theta \geqslant \frac{9}{10}$.

3.5   A uniform rod XY, of length $2a$ and weight W, is hinged to a vertical post at X. It is supported in a horizontal position by a string attached at Y and to a point Z vertically above X. A weight w is hung from Y.
(a) If the reaction at the hinge is at 90° to YZ, prove that the length of the string
   YZ is $2a \sqrt{\dfrac{2(W + w)}{W}}$.
(b) Find the tension in the string.   (SUJB)

3.6   A uniform rod AB, of length $2a$ and weight W, is hinged to a vertical wall at A and is supported at an angle θ to the horizontal (B above A) by a string of length $2a$ attached to B and to a point C on the wall vertically above A. A load of weight W is hung from B.

Find the tension in the string and the force exerted by the hinge on the rod, in terms of W and θ.

Show that, if the reaction at the hinge A is perpendicular to the string, then $AC = a\sqrt{6}$.

3.7 Two uniform cylinders of equal radius rest against each other on a fixed plane inclined at an angle α to the horizontal, the lower cylinder resting against a fixed vertical plane. The line of intersection of the planes is perpendicular to the

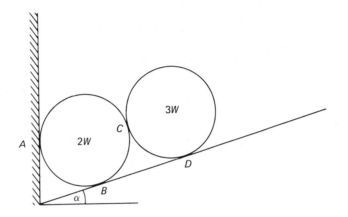

section shown. The lower cylinder has a weight $2W$ and the upper cylinder has weight $3W$. Given that all the contacts are smooth, find the magnitudes of the reactions at all points of contact, A, B, C and D.

3.8 Two smooth planes inclined at angles α, β to the horizontal intersect in a horizontal line. A non-uniform thin heavy bar AB rests with end A on the plane of inclination α and the end B on the plane of inclination β, with AB perpendicular to the line intersection of the planes. The mass of the bar is M and the mass centre at distances a and b from A and B respectively.

Prove that, if θ is the inclination of the bar to the vertical in equilibrium, $(a + b) \cot \theta = | a \cot \alpha - b \cot \beta |$ and find the reaction at A in terms of α and β.

(OLE)

3.9 The figure shows a framework $ABCDEF$ consisting of nine smoothly jointed light rods. The rods AB, BC, CD and EF are each of length 8 m and are horizontal. The rods FB and EC are each of length 6 m and are vertical. The framework is

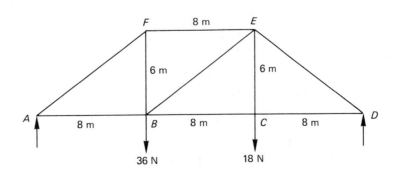

simply supported in a vertical plane at A and D. Loads of 36 N and 18 N are attached at B and C respectively. Calculate

(a) the reactions at A and D,

(b) the forces acting in AF, AB and DE,

(c) the forces acting in BF and BE.

An additional load of 18 N is now attached at C. Find the force in BE.

(AEB 1984)

3.4 Brief Solutions to Exercises

3.1 (a) Resolving gives 6 N parallel to OC, 8 N parallel to OA, so $R = 10$ N at angle $\tan^{-1} \frac{3}{4}$ to OA.

If $(0, y)$ lies on line of action then moments about O:

$3(1) - 2(1) = 8(y)$ $\Rightarrow$ $y = \frac{1}{8}$, so line of action goes through $(0, \frac{1}{8})$. Gradient is $\frac{3}{4}$, so $8y = 6x + 1$.

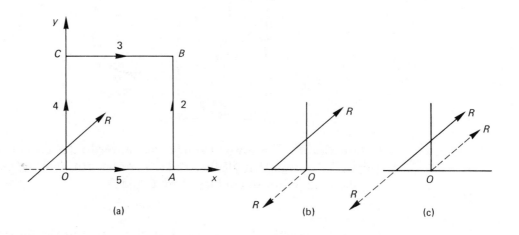

(a)   (b)   (c)

(b) Introduce force $-\mathbf{R}$ at O along $4y = 3x$. Moment of couple = 1 N m in sense $OCBA$.

(c) Add couple, force $-\mathbf{R}$ along $8y = 6x + 1$ and force $\mathbf{R}$ at O. Moment of couple = 1 N m in sense $OABC$.

Resultant is 10 N along $4y = 3x$, x increasing.

3.2 $\rightarrow$, $R_x = 1 \cos 60° + 3 + 5 \cos 60° - 6 + 2\sqrt{3} \cos 30° = 3$ N.

$\downarrow$, $R_y = -1 \cos 30° + 5 \cos 30° - 2\sqrt{3} \sin 30° = \sqrt{3}$ N.

$R = 2\sqrt{3}$ N at $\tan^{-1} \dfrac{\sqrt{3}}{3}$ with AB, i.e. $30°$ to AB on opposite side to D and C.

Taking moments about A, $3 \left(a \dfrac{\sqrt{3}}{2} \right) + 5 (a\sqrt{3}) = R_y AX$. Therefore $AX = \dfrac{13}{2} a$.

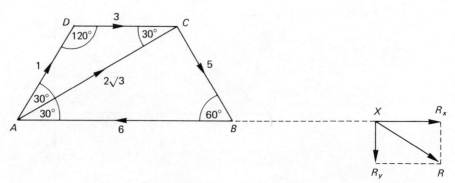

42

3.3 Taking moments about P, $\quad T(2l)\cos 60° = 60(l)\sin 60°$.

$\qquad T = 30\sqrt{3} \approx 52$. **Answer C**

3.4 Moments about A: $\quad P(3a) = S(a) + W(2a)\cos\theta + 2W(4a)\cos\theta$.

Resolving vertically, $\quad 3W + S\cos\theta = P\cos\theta$.

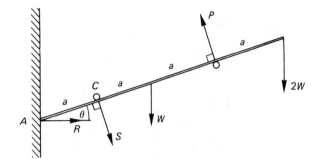

Solving, $\quad S = W\dfrac{(10\cos^2\theta - 9)}{2\cos\theta}$, $\quad P = W\dfrac{(10\cos^2\theta - 3)}{2\cos\theta}$.

But $\quad P \geqslant 0$, $\Rightarrow 10\cos^2\theta \geqslant 3$; $\quad S \geqslant 0$, $\Rightarrow 10\cos^2\theta \geqslant 9$. So $\cos^2\theta \geqslant \dfrac{9}{10}$.

3.5 (a) For equilibrium all forces must pass through N $\Rightarrow$ resultant of W and w acts at M.

$$NY = 2a\cos\theta, \quad MY = \frac{Wa}{w + W}, \quad \cos\theta = \frac{MY}{NY}, \quad \Rightarrow \quad \cos^2\theta = \frac{W}{2(w + W)}.$$

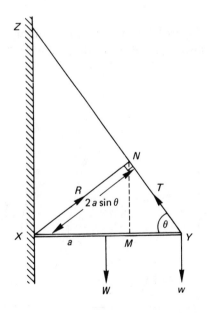

Alternatively, take moments about N.

$$\text{Length of string} = \frac{2a}{\cos\theta} = 2a\sqrt{\left(\frac{2(w + W)}{W}\right)}.$$

(b) Taking moments about X: $\quad T(2a\sin\theta) = Wa + w(2a)$,

$$T = \frac{2w + W}{2\sin\theta} = \sqrt{\left(\frac{(2w + W)(w + W)}{2}\right)}.$$

43

3.6 Taking moments about A: $T(2a \sin 2\theta) = Wa \cos \theta + W(2a \cos \theta)$,

$$T = \frac{3W}{4 \sin \theta}.$$

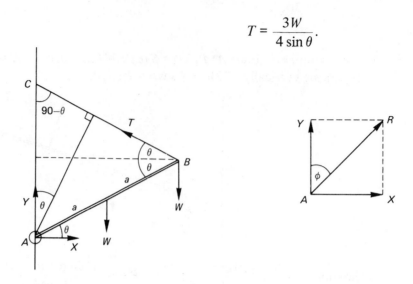

Resolving, $X = T \cos \theta = \dfrac{3W \cot \theta}{4}$, $Y = 2W - T \sin \theta = \dfrac{5}{4}W$.

Reaction $R = \dfrac{W}{4} \sqrt{(9 \cot^2 \theta + 25)}$ at $\tan^{-1} \dfrac{X}{Y} = \tan^{-1} \left(\dfrac{3 \cot \theta}{5} \right)$

$= \phi$ to the vertical.

If R is perpendicular to BC, $\phi = \theta$, so $\tan^2 \theta = \frac{3}{5}$.

$AC = 2(2a \sin \theta)$ $\Rightarrow$ $AC = a\sqrt{6}$.

3.7   Resolving: for upper cylinder, $R_4 = 3W \cos \alpha$, $R_3 = 3W \sin \alpha$;
for lower cylinder, $2W \sin \alpha + R_3 = R_1 \cos \alpha$ $\Rightarrow$ $R_1 = 5W \tan \alpha$,

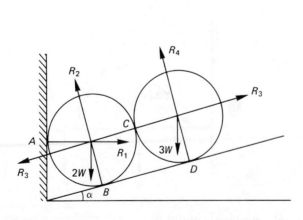

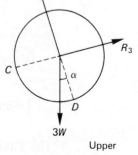

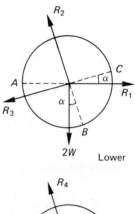

$$2W \cos \alpha + R_1 \sin \alpha = R_2 \quad \Rightarrow \quad R_2 = \frac{W(2 \cos^2 \alpha + 5 \sin^2 \alpha)}{\cos \alpha}.$$

3.8 Assume $\alpha > \beta$. Resolving vertically, $R_A \cos\alpha + R_B \cos\beta = Mg$;

horizontally, $R_A \sin\alpha = R_B \sin\beta$.

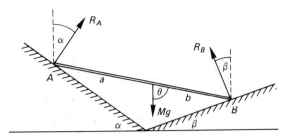

So $R_B = R_A \dfrac{\sin\alpha}{\sin\beta}$ and $R_A = \dfrac{Mg}{(\cos\alpha + \sin\alpha \cot\beta)} = \dfrac{Mg \sin\beta}{\sin(\alpha + \beta)}.$

Taking moments about B:

$$R_A \sin\alpha (a + b) \cos\theta + R_A \cos\alpha (a + b) \sin\theta = Mgb \sin\theta.$$

Substituting for R_A gives: $(a + b) \cot\theta = b \cot\beta - a \cot\alpha.$

If $\alpha < \beta$, $(a + b) \cot\theta = a \cot\alpha - b \cot\beta$. Hence the use of modulus signs.

Alternatively, all three forces must pass through one point, so, by Lami's theorem:

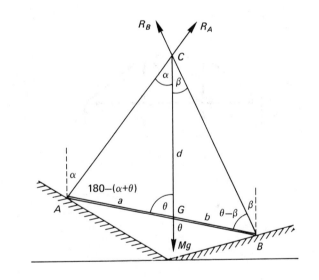

$$\frac{Mg}{\sin(\alpha + \beta)} = \frac{R_A}{\sin\beta} = \frac{R_B}{\sin\alpha} \;\Rightarrow\; R_A = \frac{Mg \sin\beta}{\sin(\alpha + \beta)}.$$

By the sine rule

$$\frac{a}{\sin\alpha} = \frac{CG}{\sin(\alpha + \theta)}, \qquad \frac{b}{\sin\beta} = \frac{CG}{\sin(\theta - \beta)}.$$

Eliminating CG gives result.

3.9 (a) Resolving vertically: $R_A + R_D = 36 + 18 = 54$ N.

Taking moments about A: $36(8) + 18(16) = R_D(24)$; $R_D = 24$ N;

$R_A = 30$ N.

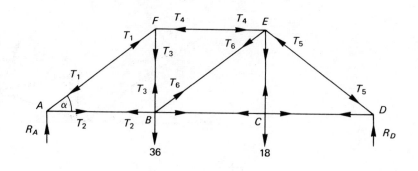

(b) Resolving: At A: $T_1 \sin\alpha = R_A$ $\Rightarrow$ $T_1 = 50$ N (compression),

$T_2 = T_1 \cos\alpha$ $\Rightarrow$ $T_2 = 40$ N (tension).

At D: $T_5 \sin\alpha = R_D$ $\Rightarrow$ $T_5 = 40$ N (compression).

(c) Resolving: At F: $T_3 = T_1 \sin\alpha$ $\Rightarrow$ $T_3 = 30$ N (tension),

$T_4 = T_1 \cos\alpha$ $\Rightarrow$ $T_4 = 40$ N (compression).

At E: $T_5 \cos\alpha + T_6 \cos\alpha = T_4$ $\Rightarrow$ $T_6 = 10$ N (tension).

Force of 18 N added at C makes $R_A = R_D = 36$ N.

By symmetry, $T_6 = 0$ $\Rightarrow$ zero force in BE.

Alternatively, use Bow's notation.

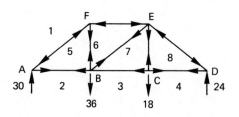

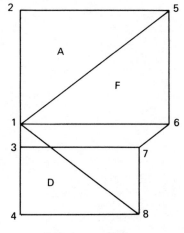

Scale 1 cm to 10 N

4 Friction. Equilibrium of Bodies in Contact

The laws of friction. Equilibrium of systems of particles and rigid bodies under the action of coplanar forces including friction.

4.1 Fact Sheet

(a) Laws of Friction

(i) Friction F opposes the movement of an object across a rough surface.
(ii) Up to a limiting value the magnitude of the frictional forces is just sufficient to prevent motion.
(iii) If μ is the coefficient of friction, and R is the normal reaction, the limiting value of the frictional force is μR; so at all times $F \leqslant \mu R$.

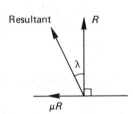

The resultant of R and μR, the limiting value of the frictional force, makes an angle λ with the normal where $\mu = \tan \lambda$. λ is called the angle of friction.

(b) Equilibrium

A system is in equilibrium when the sum of all the forces = 0 *and* the sum of moments about any point = 0.
  When two or more bodies in contact are in equilibrium then
(i) the complete system is in equilibrium, and
(ii) each body is in equilibrium.

(c) **Hints**

In general:

(i) Good clear diagrams showing the directions of all the forces are essential.
(ii) Questions which state 'on the point of moving', 'limiting equilibrium', 'will just prevent motion' and 'will just move' all require the use of $F = \mu R$.
(iii) A body will slide if the force F required to overcome the other forces is such that $F \geqslant \mu R$.
(iv) A body will topple, rotate, turn or tilt about a point A if the turning moment about A (in the direction of toppling) $\geqslant 0$ before $F = \mu R$.
(v) Look for the equations which involve the least number of variables first.

For systems of bodies:

(i) Treat as a whole and as separate bodies, with a clear diagram for each.
(ii) Look for symmetry of configuration. This can reduce the number of unknowns considerably.
(iii) For the whole system the internal forces are in equilibrium. At each point of contact between bodies the reactions are equal and opposite.
(iv) Avoid writing down all possible equations. The number of independent equations needed is equal to the number of unknowns.

4.2 Worked Examples

4.1 A particle of mass m rests on a rough horizontal plane and is pulled by a force of magnitude $(mg)/\sqrt{3}$ inclined at an angle $60°$ to the horizontal. Find the minimum value of μ, the coefficient of friction between the particle and the plane if the particle does not move.

● Let the normal reaction and frictional force be R and F respectively.

Resolving vertically, $\quad R + \dfrac{mg}{\sqrt{3}} \sin 60° = mg$,

$$R = mg - \frac{mg}{2} = \frac{mg}{2} \tag{1}$$

Resolving horizontally, $\quad F = \dfrac{mg}{\sqrt{3}} \cos 60° = \dfrac{mg}{2\sqrt{3}}$. $\tag{2}$

Since the particle does not move, $\quad F \leqslant \mu R$,

From (1) and (2) $\qquad \dfrac{F}{R} = \dfrac{1}{\sqrt{3}} \quad \Rightarrow \quad \mu \geqslant \dfrac{1}{\sqrt{3}}$.

Therefore the minimum value of μ is $\dfrac{1}{\sqrt{3}}$.

4.2 A uniform rod AB, of length $2l$ and weight W, is in equilibrium with the end A on a rough horizontal floor and the end B against a smooth vertical wall. The rod makes an angle $\tan^{-1} 2$ with the horizontal and is in a vertical plane which is perpendicular to the wall. Find the least possible value of μ, the coefficient of friction between the floor and the rod.

Given that $\mu = \frac{5}{16}$, find the distance from A of the highest point of the rod at which a particle of weight W can be attached without disturbing equilibrium.

(L)

- $\tan \theta = 2$.

For equilibrium:
Take the forces as shown in the diagram.
Resolving forces vertically, $R = W$.   (1)
Resolving forces horizontally, $S = F$.   (2)
Taking moments about A, $S(2l \sin \theta) = W(l \cos \theta)$.   (3)

Therefore $$S = \frac{W}{2 \tan \theta} = \frac{W}{4} = F \qquad \text{from (2)}.$$

Therefore $\frac{F}{R} = \frac{1}{4}$,   but $\frac{F}{R} \leqslant \mu$,   so $\mu \geqslant \frac{1}{4}$.

Least value of μ is $\frac{1}{4}$.

Let the particle be attached to the rod a distance x from A.
Resolving forces vertically, $R_1 = 2W$.   (4)
Resolving forces horizontally, $S_1 = F_1$.   (5)
Taking moments about A, $S_1(2l \sin \theta) = W(x \cos \theta) + W(l \cos \theta)$.   (6)

Therefore $$S_1 = \frac{W(x+l)}{2\,l \tan \theta} = \frac{W(x+l)}{4l} = F_1 \qquad \text{from (5)}.$$

Therefore $\dfrac{F_1}{R_1} = \dfrac{(x+l)}{2(4l)} = \dfrac{1}{8}\left(\dfrac{x}{l} + 1\right)$.

Given $\mu = \frac{5}{16}$, $\frac{F_1}{R_1} \leqslant \frac{5}{16}$,   therefore $\frac{x}{l} + 1 \leqslant \frac{5}{2}$, $\frac{x}{l} \leqslant \frac{3}{2}$. Therefore $x \leqslant \frac{3}{2}l$.

Therefore the highest point at which the particle can be attached is $\frac{3}{2}l$ from A.

4.3   A uniform square lamina $ABCD$ of mass M rests in a vertical plane with AB in contact with a horizontal table. The coefficient of friction between the lamina and the table is μ.

A gradually increasing horizontal force P is applied at C in the plane of the lamina in the direction $\overline{DC}$. Prove that equilibrium is broken by sliding if $\mu < 0.5$, and by the lamina tilting about B if $\mu > 0.5$. Find the value of P which breaks equilibrium in the latter case.   (OLE)

- Let the lamina have side $2a$, R denote the normal reaction acting somewhere along the base and F the frictional force.
For sliding:
Resolving horizontally, $P \geqslant F = \mu R$.
Resolving vertically, $R = Mg$.
Therefore for sliding $P \geqslant \mu Mg$.
For tilting:
When about to tilt, R acts at B.
Taking moments about B, $P(2a) \geqslant Mga$.
Therefore for tilting $P \geqslant Mg/2$.
If $\mu < \frac{1}{2}$, P attains the sliding condition first.
If $\mu > \frac{1}{2}$, P attains the tilting condition first.
Value of P which breaks equilibrium by tilting is $P = Mg/2$.

49

4.4 Two uniform rods AB, BC each of length $2a$ and weight W are smoothly jointed at B and A is fixed by a smooth hinge to a rough vertical wall. The system rests in equilibrium in a vertical plane perpendicular to the wall with C in contact with the wall and each rod inclined at angle θ to the vertical. The coefficient of friction between C and the wall is μ.

(a) Find the magnitudes of the frictional force and normal reaction at C.

(b) Show that $\tan \theta \geqslant \dfrac{2}{\mu}$.

(c) A force of magnitude W, applied vertically downwards at C, is just sufficient to cause slipping. Find the value of μ.   (SUJB)

- *For the whole system:*

  Taking moments about A, $2W (a \sin \theta) = R (4a \cos \theta)$,
  $$R = (W \tan \theta)/2. \tag{1}$$

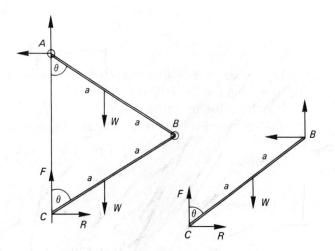

For rod CB:

Taking moments about B, $F (2a \sin \theta) = R (2a \cos \theta) + W (a \sin \theta)$,
$$F = R \cot \theta + W/2 = W. \tag{2}$$

(a) The frictional force and normal reaction at C are W and $\dfrac{W \tan \theta}{2}$ respectively.

(b) For equilibrium, $F \leqslant \mu R$,   i.e. $W \leqslant \mu \dfrac{W \tan \theta}{2}$.   Therefore $\tan \theta \geqslant \dfrac{2}{\mu}$.

(c) When the extra force W is applied at C equation (1), and hence R, remain unchanged.

For rod CB:

Let the new frictional force at C be F_1.

Taking moments about B:

$$F_1 (2a \sin \theta) - W (2a \sin \theta) = R (2a \cos \theta) + W (a \sin \theta);$$

$$F_1 - W = \frac{W}{2} + \frac{W}{2} = W \Rightarrow F_1 = 2W.$$

In limiting equilibrium, $F_1 = \mu R$,

$$2W = \mu \frac{W \tan \theta}{2},$$

$$\mu = \frac{4}{\tan \theta}.$$

4.5 An L-shaped structure is formed by rigidly joining two uniform rods AB and AC at A so that the angle BAC is a right angle. The rod AB is of mass $3m$ and length $4a$ and the rod AC is of mass $2m$ and length $2a$.

The structure is free to turn in a vertical plane about a smooth horizontal axis through A. A horizontal force is applied at C so that the structure remains in equilibrium with AB and AC equally inclined to the vertical with B and C lower than A. Determine

(a) The magnitude of the horizontal force.

(b) The magnitude of the reaction at A.

(c) The tangent of the angle that the reaction at A makes with the horizontal.

The applied force at C is now removed and the axis at A roughened, so that it exerts a frictional couple on the structure, which remains in equilibrium with AB making an acute angle θ with the upward vertical and B and C on the same side of the vertical through A.

(d) Find the value of the frictional couple necessary to maintain this equilibrium.

(e) The nature of the frictional couple is such that it will take whatever value is necessary to maintain equilibrium, up to a maximum value of G. Find the least such G to ensure that equilibrium is possible for all values of the acute angle θ. (AEB 1984)

- Let the force applied at C be F, and the reaction at A have horizontal and vertical components X and Y.

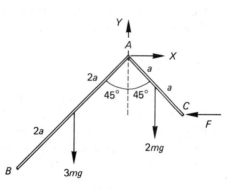

Taking moments about A,

$$3mg\,(2a\sin 45°) = 2mg\,(a\sin 45°) + F\,(2a\cos 45°),$$

$$F = 2mg.$$

(a) Magnitude of the horizontal force is $2mg$.

(b) Resolving horizontally and vertically:

$$X = F = 2mg \quad \text{and} \quad Y = 5\,mg.$$

Magnitude of the reaction at $A = mg\sqrt{(2^2 + 5^2)} = mg\sqrt{29}$.

(c) Reaction at A makes an angle with the horizontal of $\tan^{-1}\frac{5}{2} = 68.2°$.

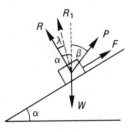

(d) Let the frictional couple about A be G.

Taking moments about A,

$$3mg\,(2a\sin\theta) + 2mg\,(a\cos\theta) = G.$$

Frictional couple is $2mga\,(3\sin\theta + \cos\theta)$

$$= 2mga\sqrt{10}\,[\sin(\theta + \alpha)] \quad \text{where } \tan\alpha = \tfrac{1}{3}.$$

(e) The least value of G to ensure that equilibrium is possible for all θ is $2mga\sqrt{10}$.

4.6 A weight W is placed on a rough plane inclined at an angle α to the horizontal and the angle of friction between the weight and the plane is $\lambda\ (<\alpha)$.

(a) Show that the weight will slide down the plane.

(b) If the weight is just prevented from sliding down by a force P inclined to the upward vertical at an acute angle β, show that $P = \dfrac{W\sin(\alpha - \lambda)}{\sin(\alpha - \lambda + \beta)}$.

Obtain the least value of P as β varies and the corresponding value of β.

(c) If $\alpha + \lambda < \dfrac{\pi}{2}$, what is the direction and magnitude of the smallest force which will just move the weight up the plane? (SUJB)

● (a) Let F be the force required for equilibrium.

Resolving normal to the plane, $R = W\cos\alpha$.

Resolving parallel to the plane, $F = W\sin\alpha$.

$$\frac{F}{R} = \tan\alpha \quad \Rightarrow \quad F = R\tan\alpha > R\tan\lambda \quad (\text{since } \alpha > \lambda).$$

Therefore $F > \mu R$. But maximum frictional force $= \mu R$.

Thus the weight will slide down the plane.

(b) When the weight is just prevented from sliding down the plane the maximum possible friction F acts up the plane

The resultant of R and F is R_1 acting at an angle λ to the normal.

Resolving vertically, $R_1\cos(\alpha - \lambda) + P\cos\beta = W$

Resolving horizontally, $R_1\sin(\alpha - \lambda) = P\sin\beta$,

$$\Rightarrow \quad \frac{P\sin\beta\cos(\alpha - \lambda)}{\sin(\alpha - \lambda)} + P\cos\beta = W.$$

Multiply by $\sin(\alpha - \lambda)$:

$$P\,[\sin\beta\cos(\alpha - \lambda) + \cos\beta\sin(\alpha - \lambda)] = W\sin(\alpha - \lambda);$$

$$P\sin(\alpha - \lambda + \beta) = W\sin(\alpha - \lambda) \quad \text{or} \quad P = \frac{W\sin(\alpha - \lambda)}{\sin(\alpha - \lambda + \beta)}.$$

As β varies, $P_{\min}$ occurs when $\alpha - \lambda + \beta = \dfrac{\pi}{2}$,

$$P_{\min} = W\sin(\alpha - \lambda) \quad \text{when} \quad \beta = \frac{\pi}{2} + \lambda - \alpha.$$

(c) When the weight is about to move up the plane the direction of F is reversed and then the resultant of R and F will make an angle $-\lambda$ with the normal.

Compared with (b): $P = \dfrac{W\sin(\alpha + \lambda)}{\sin(\alpha + \lambda + \beta)}$.

$$P_{\min} = W\sin(\alpha + \lambda) \quad \text{when} \quad \beta = \frac{\pi}{2} - (\alpha + \lambda).$$

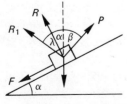

4.7 Two uniform rods AB, BC of equal lengths are freely joined together at B. The mass of AB is m and that of BC is $2m$. The rods stand in equilibrium in a vertical plane on a rough horizontal plane, and the angle $ABC = 2\alpha$.

The coefficient of friction at A is μ_1, and the coefficient of friction at C is μ_2. Find the horizontal and vertical components of the reaction between the rods at B and prove that $\mu_1 \geqslant \frac{3}{5} \tan \alpha$ and $\mu_2 \geqslant \frac{3}{7} \tan \alpha$. (OLE)

● Take the forces as shown in the diagram.

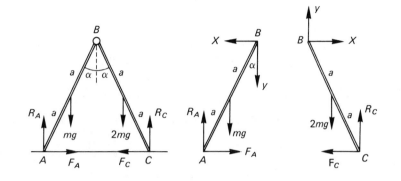

For rod AB, taking moments about A,

$$X (2a \cos \alpha) = Y (2a \sin \alpha) + mg (a \sin \alpha) \quad \Rightarrow \quad X = \left(Y + \frac{mg}{2}\right) \tan \alpha. \quad (1)$$

For rod BC, taking moments about C,

$$X (2a \cos \alpha) + Y (2a \sin \alpha) = 2mg (a \sin \alpha) \quad \Rightarrow \quad X = (mg - Y) \tan \alpha. \quad (2)$$

Add (1) and (2): $2X = \dfrac{3mg}{2} \tan \alpha, \quad X = \dfrac{3mg}{4} \tan \alpha.$

Subtract (2) from (1): $0 = \left(2Y - \dfrac{mg}{2}\right) \tan \alpha, \quad Y = \dfrac{mg}{4}.$

Thus the horizontal and vertical components of the reaction at B are $\frac{3}{4}mg \tan \alpha$ and $\frac{1}{4}mg$.

Resolving horizontally for each rod, $X = F_A = F_C = \dfrac{3mg}{4} \tan \alpha.$

Resolving vertically for each rod, $R_A = mg + Y = \dfrac{5mg}{4}$,

$$R_C + Y = 2mg \quad \Rightarrow \quad R_C = \frac{7mg}{4}.$$

At A, $\dfrac{F_A}{R_A} = \dfrac{3mg \tan \alpha}{5mg} = \frac{3}{5} \tan \alpha.$ At C, $\dfrac{F_C}{R_C} = \dfrac{3mg \tan \alpha}{7mg} = \frac{3}{7} \tan \alpha.$

But, since the system is in equilibrium, $\mu_1 \geqslant \dfrac{F_A}{R_A}, \quad \mu_2 \geqslant \dfrac{F_C}{R_C},$

so $\mu_1 \geqslant \frac{3}{5} \tan \alpha, \quad \mu_2 \geqslant \frac{3}{7} \tan \alpha.$

4.8 A uniform circular hoop of weight W hangs over a rough horizontal peg A. The hoop is pulled with a gradually increasing horizontal force P which is applied at the other end B of the diameter through A and acts in the vertical plane of the hoop.

Given that the system is in equilibrium and that the hoop has not slipped when AB is inclined at an angle θ to the downward vertical, find the value of P in terms of W and θ. Show also that the ratio of the frictional force to the normal reaction at the peg is

$$(\tan \theta)/(2 + \tan^2 \theta).$$

Show that, when the coefficient of friction is 0.5, the hoop never slips, however hard it is pulled. (L)

- The forces acting on the hoop are as shown in the sketch.
  If the system is in equilibrium then the turning moment about A is zero.

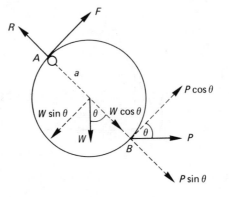

$$W(a \sin \theta) = P(2a \cos \theta) \qquad \Rightarrow \qquad P = \frac{W}{2} \tan \theta.$$

Resolving radially, $\quad R = W \cos \theta + P \sin \theta = \dfrac{W}{2}(2 \cos \theta + \sin \theta \tan \theta)$.

Resolving tangentially, $\quad F = W \sin \theta - P \cos \theta = W \sin \theta - \dfrac{W}{2} \sin \theta = \dfrac{W}{2} \sin \theta$;

$$\frac{F}{R} = \frac{\sin \theta}{(2 \cos \theta + \sin \theta \tan \theta)} = \frac{\tan \theta}{(2 + \tan^2 \theta)}.$$

This may be written: $\quad \dfrac{F}{R} = \dfrac{\sin \theta \cos \theta}{2 \cos^2 \theta + \sin^2 \theta} = \dfrac{\sin 2\theta}{2(1 + \cos^2 \theta)}.$

Now $\sin 2\theta \leqslant 1$ for all θ and $1 + \cos^2 \theta \geqslant 1$ for all θ.

Since $\sin 2\theta = 1$ and $1 + \cos^2 \theta = 1$ occur at different values of θ, $\dfrac{F}{R} < \tfrac{1}{2}$ for all θ.

Hence if $\mu = \tfrac{1}{2}$ the hoop never slips.

4.9 $ABCD$ is a light inextensible string of length $38a$ with $AB = 15a$ and $BC = 10a$. A and D are fixed to small rings, each of mass 2 kg, which can slide on a rough rail which is fixed horizontally.

Masses W and W' are attached to B and C respectively and the system rests in equilibrium with BC horizontal and distant $12a$ below AD. Show that $9W = 5W'$. If the coefficient of friction between each ring and the rail is 0.25 and one of the rings is on the point of slipping, determine which it is and the value of W.

(SUJB)

- Let $\angle DAB = \alpha$, $\angle ADC = \beta$. $CD = 13a$.
  $\sin \alpha = \tfrac{4}{5}$, $\sin \beta = \tfrac{12}{13}$.
  By right-angled triangles, $AN = 9a$ and $MD = 5a$.

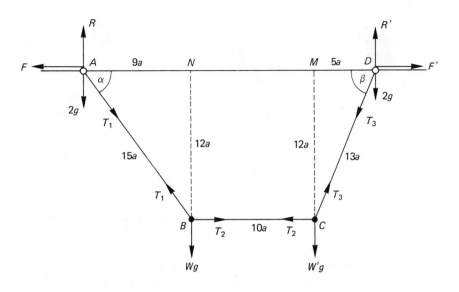

Let the tensions, reactions and frictional forces be as shown in diagram.

At B, resolving horizontally, $T_1 \cos\alpha = T_2$; (1)

resolving vertically, $T_1 \sin\alpha = Wg$. (2)

Thus $Wg = T_2 \tan\alpha = \frac{12}{9} T_2$. (3)

Similarly at C, $W'g = T_2 \tan\beta = \frac{12}{5} T_2$. (4)

Therefore $9W = 5W'$.

At A, resolving horizontally and vertically,

$$F = T_1 \cos\alpha = Wg \cot\alpha = \tfrac{3}{4} Wg \quad \text{(from (1))},$$

$$R = 2g + T_1 \sin\alpha = 2g + Wg \quad \text{(from (2))},$$

$$\frac{F}{R} = \frac{3Wg}{8g + 4Wg} = \frac{3W}{8 + 4W}.$$ (5)

Similarly at D, $R' = 2g + W'g$, $F' = W'g \cot\beta = \frac{5}{12}W'g$.

But $W' = \frac{9}{5} W$, so $R' = g(2 + \frac{9}{5} W)$ and $F' = \frac{9}{12} Wg$.

$$\frac{F'}{R'} = \frac{3W/4}{2 + 9W/5} = \frac{3W}{8 + 36W/5}.$$

Since $4W < \dfrac{36W}{5}$, $\dfrac{F}{R} > \dfrac{F'}{R'}$, so A will slip first.

Given $\dfrac{F}{R} = \frac{1}{4}$, $\dfrac{3W}{8 + 4W}$ (5) $\Rightarrow$ $12W = 8 + 4W$, $\Rightarrow$ $W = 1$.

Therefore the value of W is 1 kg.

4.10 A uniform sphere of weight W and radius a rests on a rough horizontal plane touching it at a point A. A rod of length $\dfrac{3\sqrt{3}}{2} a$ and weight $3W$ is hinged to the plane at a point B and rests against the sphere, the point of contact being C.

Points A, B, and C are in the same vertical plane.

If the coefficient of friction at A is μ, and at C is 2μ, and angle $ABC = 60°$ calculate

(a) the normal reactions at A and C,

(b) the horizontal and vertical components of the reaction at the hinge at B,

(c) the smallest value of μ for which this equilibrium position is possible.

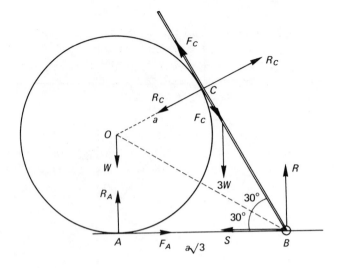

● See the notation in the figure.

In triangle OAB, $B = 30°$, so $AB (= BC) = a\sqrt{3}$.

For the whole system:

Since F_C and R_C are internal forces (i.e. are in equal and opposite pairs) they will make no contribution when taking moments or resolving forces in any direction.

Resolving horizontally, $F_A = S$.

Taking moments about B, $R_A (a\sqrt{3}) = W (a\sqrt{3}) + 3W \left(\dfrac{3\sqrt{3}}{4} a \cos 60°\right)$.

$$R_A = W \left(1 + \tfrac{9}{8}\right) = \tfrac{17}{8} W. \tag{1}$$

Resolving vertically, $R_A + R = 4W$ $\Rightarrow$ $R = \tfrac{15}{8} W.$ $\tag{2}$

For the rod:

Taking moments about B, $R_C (a\sqrt{3}) = 3W \left(\dfrac{3\sqrt{3}}{4} a \cos 60°\right)$

$$\Rightarrow R_C = \tfrac{9}{8} W. \tag{3}$$

Normal reactions at A and C are $\tfrac{17}{8} W$ and $\tfrac{9}{8} W$ respectively.

For the sphere:

Taking moments about C, $F_A (a + a \cos 60°) + W (a \sin 60°) = R_A (a \sin 60°)$;

$$\tfrac{3}{2} F_A = \tfrac{17}{8} \tfrac{\sqrt{3}}{2} W - \tfrac{\sqrt{3}}{2} W = \tfrac{9\sqrt{3}}{16} W \quad \Rightarrow \quad F_A = \tfrac{3\sqrt{3}}{8} W = S.$$

Horizontal and vertical components of the reaction at the hinge are $\dfrac{3\sqrt{3}}{8} W$ and $\dfrac{15}{8} W$ respectively.

Taking moments about O, $F_A = F_C$.

At A, $\dfrac{F_A}{R_A} = \dfrac{3\sqrt{3}}{17} \leqslant \mu$; at C, $\dfrac{F_C}{R_C} = \dfrac{\sqrt{3}}{3} \leqslant 2\mu.$

Thus $\mu \geqslant \dfrac{3\sqrt{3}}{17}$ and $\mu \geqslant \dfrac{\sqrt{3}}{6}$, hence $\mu \geqslant \dfrac{3\sqrt{3}}{17}$.

4.3 Exercises

4.1 A particle of weight W is placed on a rough plane which is inclined at an angle α to the horizontal. The coefficient of friction between the particle and the plane is μ. Show that if $\mu < \tan \alpha$ then the particle will slide down the plane.

If $\alpha = 45°$ and $\mu = 0.5$, find the magnitude of the least horizontal force needed to maintain the particle in equilibrium.

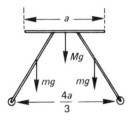

4.2 A particle is placed on the inner surface of a fixed rough hollow sphere of internal radius a. Given that the coefficient of friction between the particle and the sphere is $\frac{3}{4}$, show that the particle rests in limiting equilibrium at a depth $4a/5$ below the centre of the sphere. (L)

4.3 The figure represents a vertical central cross-section $ABCD$ of a uniform cube of side $2a$ and weight W resting on a rough horizontal plane. The coefficient of friction between the cube and the plane is μ. A force of magnitude P and inclined at an acute angle θ ($> 45°$) to the downward vertical is applied at A, in the plane $ABCD$.

Find the value of P when
(a) the cube is about to tilt
(b) the cube is about to slide.
Hence determine the range of values of μ such that, as P is gradually increased, the cube slips before it tilts.

The direction of the force at A is now changed so that while remaining at an angle θ to the downward vertical, its positive horizontal component is towards the inside of the cube, the positive vertical component still being downward.

Determine the value of P when the cube is about to tilt, provided μ is sufficiently large. (AEB 1984)

4.4 The figure shows two uniform planks, each of length a and mass m, which are freely hinged to the ground at points distance $4a/3$ apart. A third plank, of length a and mass M, rests horizontally at its points of trisection on the other two planks which are equally inclined upwards and towards each other. Show that, if μ is the coefficient of friction between the planks, then equilibrium is possible provided that

$$\mu \geqslant (m + M)/(M\sqrt{3}).$$

Show further that, if m remains constant while M increases, then the smallest value of μ consistent with equilibrium decreases, but this value must always be greater than $1/\sqrt{3}$. (L)

4.5 Two rough wires, OA and OB, are fastened together at O at right angles and are placed so that OA is vertical, OB is horizontal and A is below O. A small ring of weight W_1 slides on OA and a small ring of weight W_2 slides on OB. The rings are joined by a taut inextensible string of length a.

Given that the coefficient of friction between each ring and the wire is μ, prove that the inclination of the string to the horizontal when both rings are in limiting equilibrium is

$$\tan^{-1} \frac{W_1 - \mu^2 W_2}{\mu(W_1 + W_2)}.$$

If the two weights are equal and $\mu = 1/2$, show that this inclination is $\tan^{-1}(3/4)$ and hence, in this case, that the tension in the string is W_1. (SUJB)

4.6 A uniform rod AB has weight W and length $2a$. It rests in equilibrium with A in contact with a rough vertical wall, B being attached to one end of a light string whose other end is fixed to a point O on the wall vertically above A; the plane OAB being perpendicular to the wall.

If angle $AOB = 30°$ and angle $OAB = 120°$, show that the coefficient of friction μ between the rod and the wall must satisfy $\mu \geqslant 1/\sqrt{3}$.

If $\mu = 1/\sqrt{3}$ and the string is shortened so that angle $AOB = \theta$, where $\theta > 30°$, describe what will happen.

4.7 A uniform ladder of length $2a$ rests in limiting equilibrium with its lower end on rough horizontal ground and its upper end against a smooth vertical wall, the vertical plane containing the ladder being perpendicular to the wall. The ladder makes an angle of $60°$ with the ground. Show that the coefficient of friction is $(\sqrt{3})/6$.

The ladder is lowered in its vertical plane while still resting against the smooth wall and the ground to make an angle of $30°$ with the ground. The coefficient of friction between the ladder and the ground remains at $(\sqrt{3})/6$. A man whose weight is four times that of the ladder starts climbing up the ladder.

Find how far he can climb up the ladder before it slips. (L)

4.8 A uniform sphere, radius a, weight W, stands in a vertical plane upon a rough horizontal floor with a point A of its circumference in contact with an equally rough vertical wall. Weights are added to the point A of the sphere until the sphere is on the point of slipping. If the added weight is $3W/2$, find the coefficient of friction.

4.9 A uniform rod AB of mass m and length $2l$ is supported symmetrically by two smooth pegs C and D, which are fixed at the same height at distance $2a$ apart. Peg C is nearer to the end A of the rod. A gradually increasing force Q acts on the

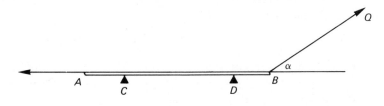

rod at B, making an angle $\alpha \left(< \dfrac{\pi}{2} \right)$ upwards from AB produced. This forces lies in the vertical plane containing AB, and the rod is kept in equilibrium by a taut string with breaking tension P attached to A in continuation of BA. Prove that the reaction on the rod at C is

$$\frac{1}{2}mg + \frac{Q(l-a)\sin\alpha}{2a},$$

and show that equilibrium is broken by the rod turning before the string breaks if

$$P > \frac{mga}{(l+a)\tan\alpha}. \tag{OLE}$$

4.4 Brief Solutions to Exercises

4.1 *For equilibrium:*
Resolving parallel to plane, $\qquad F = W \sin\alpha$.
Resolving perpendicular to plane, $\quad R = W \cos\alpha$.
So maximum frictional force $= \mu R = \mu W \cos\alpha$.
Particle slides if $\quad W \sin\alpha > \mu W \cos\alpha \quad \Rightarrow \quad \mu < \tan\alpha$.

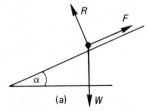

(a)

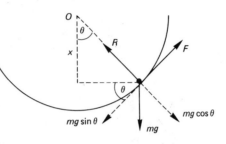

If horizontal force P *is applied:*

Resolving perpendicular to plane, $\quad R_1 = \dfrac{(W + P)}{\sqrt{2}}$.

Resolving parallel to plane, $\qquad F_1 = \dfrac{(W - P)}{\sqrt{2}}$.

P is minimum when $F_1 = \dfrac{R_1}{2} \quad \Rightarrow \quad P = \dfrac{W}{3}$.

4.2 Limiting equilibrium $\quad \Rightarrow \quad F = \dfrac{3}{4}R$.

If R makes angle θ with vertical, $\quad F = mg \sin\theta, \; R = mg \cos\theta \quad \Rightarrow \quad \tan\theta = \dfrac{F}{R} = \dfrac{3}{4}$.

Depth below centre, $x = a \cos\theta = \dfrac{4}{5}a$.

4.3 (a) When about to tilt, reaction acts at D.

Moments about D: $\quad Wa = P(2a \sin\theta)$,

$$P = \frac{W}{2 \sin\theta}.$$

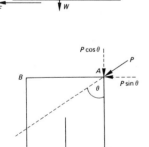

(b) When about to slide, $\quad F = \mu R \quad \Rightarrow \quad P \sin\theta = \mu(W + P \cos\theta)$

$$\Rightarrow \quad P = \frac{\mu W}{\sin\theta - \mu \cos\theta}.$$

Slips before tilts if $\quad \dfrac{\mu W}{\sin\theta - \mu \cos\theta} < \dfrac{W}{2 \sin\theta}$.

Range for μ is $\quad 0 \leqslant \mu < \dfrac{\sin\theta}{2 \sin\theta + \cos\theta}$.

If direction of P is changed, the cube will tilt about C

if $\quad P(2a \sin\theta) \geqslant Wa + P(2a \cos\theta); \quad P_{min} = \dfrac{W}{2(\sin\theta - \cos\theta)}$.

59

4.4 *For the whole system:*
Resolving vertically, $Y = g(m + M/2)$.

For AB, resolving, $R + mg = Y \Rightarrow R = \dfrac{Mg}{2}$;

   moments about $A \Rightarrow X = \dfrac{g(m + M)}{2\sqrt{3}}$.

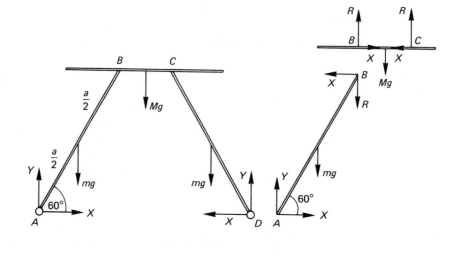

Equilibrium if $\dfrac{X}{R} \leqslant \mu \Rightarrow \mu \geqslant \dfrac{m + M}{M\sqrt{3}} = \dfrac{1}{\sqrt{3}}\left(1 + \dfrac{m}{M}\right)$.

As M increases, μ decreases but is always greater than $\dfrac{1}{\sqrt{3}}$.

4.5   In limiting equilibrium, frictional forces are μR_1 and μR_2.
For W_1, $R_1 = T\cos\alpha$, $\mu R_1 + T\sin\alpha = W_1 \Rightarrow T(\mu\cos\alpha + \sin\alpha) = W_1$.
For W_2, $\mu R_2 = T\cos\alpha$, $R_2 = W_2 + T\sin\alpha \Rightarrow T(\cos\alpha - \mu\sin\alpha) = \mu W_2$.

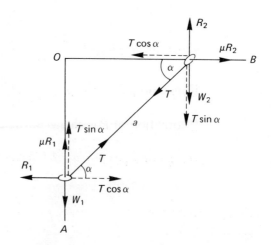

Dividing gives $\tan\alpha = \dfrac{W_1 - \mu^2 W_2}{\mu(W_1 + W_2)}$.

If $W_1 = W_2$, $\mu = \frac{1}{2}$,   then $\tan\alpha = \frac{3}{4}$.
Substituting gives $T = W_1$.

4.6 Taking moments about A gives $T = \dfrac{W\sqrt{3}}{2}$.

Resolving horizontally gives $R = T \sin 30 = \dfrac{W\sqrt{3}}{4}$.

Resolving vertically gives $F = W - T \cos 30 = \dfrac{W}{4}$.

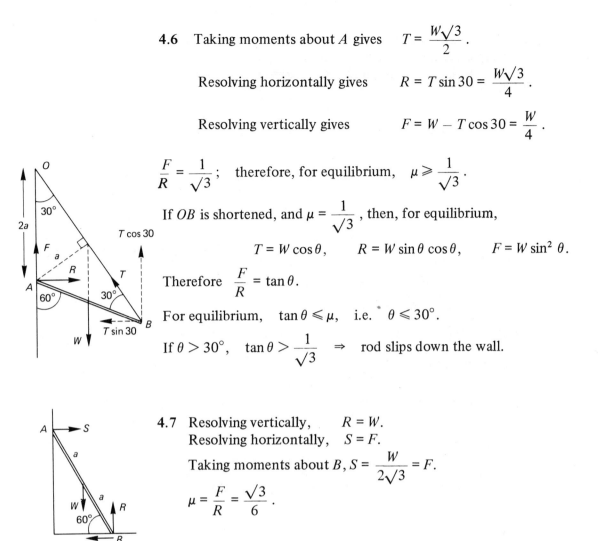

$\dfrac{F}{R} = \dfrac{1}{\sqrt{3}}$; therefore, for equilibrium, $\mu \geqslant \dfrac{1}{\sqrt{3}}$.

If OB is shortened, and $\mu = \dfrac{1}{\sqrt{3}}$, then, for equilibrium,

$$T = W \cos\theta, \qquad R = W \sin\theta \cos\theta, \qquad F = W \sin^2\theta.$$

Therefore $\dfrac{F}{R} = \tan\theta$.

For equilibrium, $\tan\theta \leqslant \mu$, i.e. $\theta \leqslant 30°$.

If $\theta > 30°$, $\tan\theta > \dfrac{1}{\sqrt{3}}$ $\Rightarrow$ rod slips down the wall.

4.7 Resolving vertically, $R = W$.
Resolving horizontally, $S = F$.

Taking moments about B, $S = \dfrac{W}{2\sqrt{3}} = F$.

$\mu = \dfrac{F}{R} = \dfrac{\sqrt{3}}{6}$.

(a)

If man is distance x up the ladder, resolving gives $R_1 = 5W$,

$$F_1 = \mu R_1 = \dfrac{5\sqrt{3}}{6} W = S_1.$$

Taking moments about B gives $S_1 a = Wa \dfrac{\sqrt{3}}{2} + 4Wx \dfrac{\sqrt{3}}{2}$

$$\Rightarrow \quad \dfrac{2}{3} Wa = 4Wx, \quad x = \dfrac{a}{6}.$$

(b)

4.8 Resolving horizontally, $R = \mu S$;

vertically, $\mu R + S = \tfrac{5}{2} W$, giving $S = \dfrac{5W}{2(\mu^2 + 1)}$.

Taking moments about A, $W = S(1 - \mu)$.
Eliminating S gives $2\mu^2 + 5\mu - 3 = 0$ $\Rightarrow$ $\mu = \tfrac{1}{2}$.

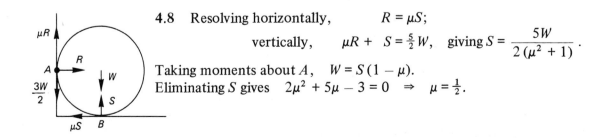

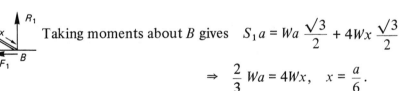

4.9   Resolving vertically, $R_C + R_D + Q \sin \alpha = mg.$
Resolving horizontally, $Q \cos \alpha = T.$
Taking moments about D, $R_C (2a) = mga + Q \sin \alpha (l - a),$

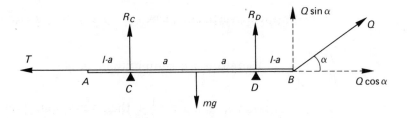

giving $R_C = \dfrac{mg}{2} + \dfrac{Q(l-a)\sin\alpha}{2a}$ and $R_D = \dfrac{mg}{2} - \dfrac{Q(l+a)\sin\alpha}{2a}$.

From these, $Q = \dfrac{mga - 2aR_D}{(l+a)\sin\alpha}$, $T = \dfrac{mga - 2R_D a}{(l+a)\tan\alpha}$.

Rod will tilt before the string breaks if $R_D = 0$ while $T < P$.

Maximum value of $T = \dfrac{mga}{(l+a)\tan\alpha}$.

Rod will turn if $P > \dfrac{mga}{(l+a)\tan\alpha}$.

5 Composite Centres of Mass

Centre of mass of composite and of truncated bodies.

5.1 Fact Sheet

Centres of mass and centres of gravity are different names for the same point.

(a) If the masses m_i and distances from the y-axis of the centres of mass $\bar{x}_i$ of two or more bodies are known then the distance of the centre of mass $\bar{x}$ of the compound body from the y-axis is given by $\bar{x} \, \Sigma \, m_i = \Sigma \, (m_i \, \bar{x}_i)$,

i.e. the moment of the whole = the sum of the moments of the parts.

(b) If parts of a body are removed then

$$\text{the moment of the remainder} = \text{the moment of the whole} - \text{the sum of the moments of the removed parts.}$$

(c) It is not necessary to use the actual masses of the parts. It is often simpler to use the ratios of the masses, especially with similar figures;

e.g.,   a cylinder of height h and radius r has mass $\pi r^2 h \rho_1$;

a cone of height h and base radius r has mass $\dfrac{\pi}{3} r^2 h \rho_2$.

These masses could be represented as $3k\rho_1$ and $k\rho_2$.

(d) If more than two parts are involved in a compound body it is convenient to tabulate the information of mass, centre of mass and moments:

Body	Mass	Centre of mass	Moments
	m_i	$(\bar{x}_i, \bar{y}_i)$	$(m_i \, \bar{x}_i, m_i \, \bar{y}_i)$

(e) When a composite body is suspended from a point and an angle with the vertical is required, it is usually easier to put a line from the point of suspension to the centre of mass on the original diagram instead of trying to draw a new one, which wastes time and obscures the coordinate system in use.

5.2 Worked Examples

5.1 A uniform straight wire of length $14a$ is bent into the shape shown in the diagram. Find the distances of the centre of mass from AB and AD.

If the wire is freely suspended from B and hangs in equilibrium, find the tangent of the angle of inclination of BC to the vertical.

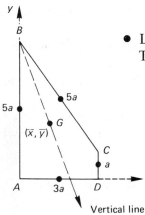

- Let the wire have a mass of $14M$.

Taking axes of reference along AD (x-axis) and AB (y-axis),

Wire	Mass	Centre of mass	Moments
AD	$3M$	$\left(\dfrac{3a}{2}, 0\right)$	$\left(\dfrac{9Ma}{2}, 0\right)$
DC	M	$\left(3a, \dfrac{a}{2}\right)$	$\left(3Ma, \dfrac{Ma}{2}\right)$
CB	$5M$	$\left(\dfrac{3a}{2}, 3a\right)$	$\left(\dfrac{15Ma}{2}, 15Ma\right)$
AB	$5M$	$\left(0, \dfrac{5a}{2}\right)$	$\left(0, \dfrac{25Ma}{2}\right)$
Total	$14M$	$(\overline{x}, \overline{y})$	$(14M\overline{x}, 14M\overline{y})$

Taking moments about the y axis,

$$\frac{9Ma}{2} + 3Ma + \frac{15Ma}{2} + 0 = 14M\overline{x},$$

$$15Ma = 14M\overline{x},$$

$$\overline{x} = \frac{15a}{14}.$$

Taking moments about the x-axis,

$$0 + \frac{Ma}{2} + 15Ma + \frac{25Ma}{2} = 14M\overline{y},$$

$$28Ma = 14M\overline{y},$$

$$\overline{y} = 2a.$$

Therefore the distances of the centre of mass from AB and AD are $\dfrac{15a}{14}$ and $2a$ respectively.

When the wire is suspended at B, the line BG will be vertical (G is the centre of mass).

The required angle is $\angle CBG = \angle ABC - \angle ABG$.

$$\tan A\hat{B}G = \frac{\overline{x}}{5a - \overline{y}} = \frac{15a/14}{3a} = \frac{5}{14}$$

$$\tan A\hat{B}C = \frac{3a}{4a} = \frac{3}{4}$$

$$\tan C\hat{B}G = \tan(A\hat{B}C - A\hat{B}G) = \frac{3/4 - 5/14}{1 + 15/56}$$

$$= \frac{42 - 20}{56 + 15}$$

$$= \frac{22}{71}.$$

The tangent of the angle of inclination of BC to the vertical is $\frac{22}{71}$.

5.2 Show by integration that the centre of mass of a uniform triangular lamina ABC is at a distance $\frac{1}{3}h$ from AB, where h is the altitude from C.

A uniform sheet of paper is in the shape of a rectangle $LMNP$, in which $LM = 8$ cm, $LP = 6$ cm. Q is a point on LM such that $LQ = 2$ cm.

The paper is cut along QN and the triangular piece is removed. Find the distance of the centre of mass of $LQNP$ from (a) LP, (b) PN.

The triangular piece QMN is then attached to the trapezium $LQNP$ so that M coincides with P, N with L, and Q lies on PN. Find the distances of the centre of mass of this composite body from LP and PN.

- Consider an elemental strip parallel to one side AB, a distance x from vertex C, and of length y, width δx.

By similar triangles, $\dfrac{y}{b} = \dfrac{x}{h}$, $y = x\dfrac{b}{h}$.

Mass of strip $= x\dfrac{b}{h}\rho\,\delta x$.

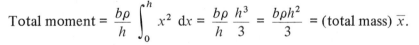

Moment about axis through C parallel to $AB = x^2\dfrac{b}{h}\rho\,\delta x$.

Total moment $= \dfrac{b\rho}{h}\displaystyle\int_0^h x^2\,\mathrm{d}x = \dfrac{b\rho}{h}\dfrac{h^3}{3} = \dfrac{b\rho h^2}{3} = $ (total mass) $\overline{x}$.

Total mass $= \frac{1}{2}bh\rho$, $\therefore$ $\overline{x} = \frac{2}{3}h$.

Therefore the centre of mass is a distance $\dfrac{h}{3}$ from AB.

Take axes of reference along PN and PL.
Centre of mass of $\triangle MQN$ is 2 cm from MQ and 2 cm from MN at $(6, 4)$.
Area of rectangle $= 48$ cm^2.
Area of triangle $= 18$ cm^2.

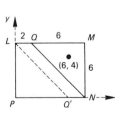

Let the mass of the rectangle be $48k$, mass of the triangle $18k$.
Then the mass of $LQNP$ is $30k$ with centre of mass $(\overline{x}, \overline{y})$.
Taking moments about PL, $\quad 30k\,(\overline{x}) + 18k\,(6) = 48k\,(4)$,

$$x = 2.8 \text{ cm.}$$

Taking moments about PN, $\quad 30k\,(\overline{y}) + 18k\,(4) = 48k\,(3)$,

$$y = 2.4 \text{ cm.}$$

(a) distance from $LP = 2.8$ cm,
(b) distance from $PN = 2.4$ cm.

When the triangle is superimposed on $LQNP$,
centre of mass of triangle is at $(2, 2)$,
centre of mass of $LQNP$ is at $(2.8, 2.4)$.
Let centre of mass of composite body be at $(\overline{x}_1, \overline{y}_1)$.
Taking moments about LP, $18k(2) + 30k(2.8) = 48k\overline{x}_1$, $\overline{x}_1 = 2.5$.
Taking moments about PN, $18k(2) + 30k(2.4) = 48k\overline{y}_1$, $\overline{y}_1 = 2.25$.
Therefore the distances of the centre of mass from LP and PN are 2.5 cm and 2.25 cm respectively.

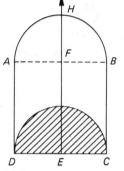

5.3　Prove that the centre of mass of a uniform lamina in the shape of a sector of a circle and subtending an angle of 2θ at its centre is distance $\dfrac{2a \sin \theta}{3\theta}$ from the centre of the circle, where a is the radius of the circle.

　　A uniform square lamina $ABCD$ of side $2a$ is made of thin card. A semicircular piece with CD as diameter is removed and attached to AB as shown in the diagram.
Show that the centre of mass of the composite lamina is at a distance $\dfrac{(4 + \pi)}{4} a$ from CD.

　　The composite lamina is now suspended from A and hangs freely in equilibrium. Show that AB is inclined to the horizontal at an angle θ, where $\tan \theta = \dfrac{4}{4 - \pi}$.

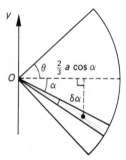

● Mass of element $= \frac{1}{2}\rho a^2\ \delta\alpha$. C. of M. is $\frac{2}{3}a$ from O $\Rightarrow$ $\frac{2}{3}a \cos \alpha$ from OY.

Total moment about $OY = \displaystyle\int_{-\theta}^{\theta} \frac{1}{3}a^3 \rho \cos \alpha \, d\alpha = \left[\frac{a^3}{3} \rho \sin \alpha \right]_{-\theta}^{\theta} = \frac{2}{3}a^3 \rho \sin \theta$.

Total mass $= \frac{1}{2}\rho a^2 (2\theta) = a^2 \rho \theta$.

Distance of centre of mass from O along line of symmetry $= \dfrac{2a^3 \rho \sin \theta}{3a^2 \rho \theta}$

$$= \dfrac{2a \sin \theta}{3\theta}.$$

Centre of mass lies on the line of symmetry EFH.
Area of square $ABCD = 4a^2$.
Area of semicircle $= \dfrac{\pi a^2}{2}$

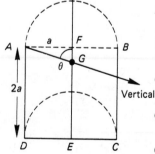

Let mass of square lamina be $4k$, then mass of semicircle $= \dfrac{\pi k}{2}$.

Centre of mass of removed semicircle is $\dfrac{4a}{3\pi}$ from E.　Put $\theta = \dfrac{\pi}{2}$ in first part.

Centre of mass of added semicircle is $\left(2a + \dfrac{4a}{3\pi} \right)$ from E.　Put $\theta = \dfrac{\pi}{2}$ in first part.

Centre of mass of composite body is $\overline{x}$ from E.

Taking moments about E, $4ka - \dfrac{\pi}{2}k\left(\dfrac{4a}{3\pi} \right) + \dfrac{\pi}{2}k\left(2a + \dfrac{4a}{3\pi} \right) = 4k\overline{x}$

$$\overline{x} = \dfrac{(4 + \pi)a}{4}.$$

Denote centre of mass by G.

Then, when the body is suspended from A, AG is vertical. If AB makes angle θ with the horizontal, it makes angle $(90 - \theta)$ with the vertical;

i.e. $\angle FAG = 90 - \theta, \quad \Rightarrow \quad \angle FGA = \theta.$

$$\tan \theta = \frac{AF}{FG}.$$

$$AF = a, \quad FG = 2a - \bar{x} = \frac{(4 - \pi)a}{4}.$$

Therefore $\tan \theta = \dfrac{4}{(4 - \pi)}.$

AB is inclined at an angle θ to the horizontal, where $\tan \theta = \dfrac{4}{4 - \pi}.$

5.4 Prove that the centre of mass of a uniform solid right circular cone of height h lies on its axis a distance $\frac{3}{4}h$ from the vertex.

The diagram represents the section of a uniform solid in the shape of a cylinder of radius 10 cm and length 12 cm joined to a cone of height 24 cm.

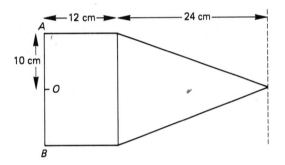

(a) Find the distance from O to the centre of mass of the solid.

(b) If the solid were freely suspended from A to hang in equilibrium under gravity, find (correct to the nearest degree) the angle that AB would make with the vertical.

(c) Show that the solid could rest in equilibrium on a horizontal table with either the curved surface of the cone or the curved surface of the cylinder in contact with the table.   (SUJB)

● Mass of a circular cone $= \frac{1}{3}\pi r^2 h\rho$.

Consider the line $y = mx$ (where $m = r/h$), rotated about the x-axis for one complete turn.

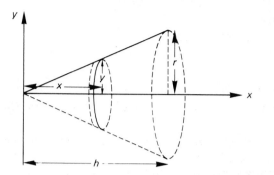

Mass of typical disc $\approx \pi y^2\, \delta x \rho$.

Moment about y-axis through $O \approx \rho\pi y^2 x\, \delta x$.

$$\text{Total moment} = \rho\pi \int_0^h y^2 x\, dx = \rho\pi m^2 \int_0^h x^3\, dx$$

$$= \left[\rho\pi\, \frac{m^2}{4}\, x^4\right]_0^h$$

$$= \frac{\rho\pi r^2}{4h^2}\, h^4 = \tfrac{1}{4}\rho\pi r^2 h^2.$$

$\tfrac{1}{3}\pi r^2 h\rho\bar{x} = \tfrac{1}{4}\rho\pi r^2 h^2 \quad\Rightarrow\quad \bar{x} = \tfrac{3}{4}h.$

Centre of mass of a cone is distance $\tfrac{3}{4}h$ from vertex.

(a) Cylinder radius 10 cm, height 12 cm;   volume $= \pi(100)\,12 = 1200\pi$.
Cone radius 10 cm, height 24 cm;   volume $= \tfrac{1}{3}\pi(100)\,24 = 800\pi$.
Mass of cylinder : mass of cone = 3 : 2.

Let masses be $3M$ and $2M$ respectively, with centres of masses at distance 6 cm and 18 cm from O, and the centre of mass of the composite solid a distance $\bar{x}$ from O.

Taking moments about AB, $3M(6) + 2M(18) = 5M(\bar{x})$

$$\bar{x} = \frac{54}{5}\text{ cm} = 10.8\text{ cm}.$$

(b) If AB makes an angle θ with the vertical then $\tan\theta = \dfrac{10.8}{10} = 1.08$,

$$\theta = 47.2°.$$

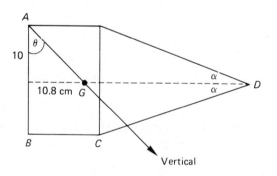

(c) (i) When the cylindrical curved surface is in contact with the table, the line of action of the weight comes within BC, so the solid can rest in equilibrium.

(ii) When the conical curved surface is in contact with the table,

$GD = 25.2$ cm; $\tan\alpha = \dfrac{10}{24} = \dfrac{5}{12} \quad\Rightarrow\quad \cos\alpha = \dfrac{12}{13}.$

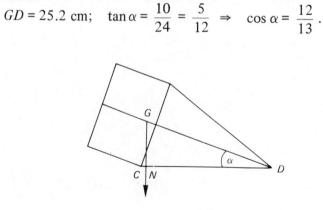

Distance $DN = GD \cos \alpha$

$\qquad = 25.2 \left(\frac{12}{13}\right) \quad = 23.3$ cm.

But $\quad CD^2 = 10^2 + 24^2, \quad CD = 26$ cm.

$DN < CD$, so line of action comes within the line CD,

and the solid can rest in equilibrium.

5.5 Prove that the centre of mass of a uniform right circular cone of semi-vertical angle α and height h is on the axis of the cone at a distance $\frac{3}{4}h$ from the vertex.

Such a cone is supported with its base area in contact with a smooth vertical wall by means of a light inelastic string joining the vertex of the cone to a point on the wall vertically above the centre of the base. Find the maximum possible length of the string. (OLE)

- (See example 5.4 for the proof.)

  Base radius $= h \tan \alpha$.

  Let the resultant reaction between the plane surface and the wall be R and the tension in the string AD be T.

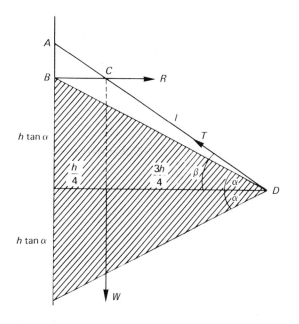

Either:

The three forces (weight, tension and reaction) must pass through one point (for equilibrium), where the line of action of the weight intersects the normal reaction R. For maximum length of the string R must act at B.

$$\tan \beta = \frac{h \tan \alpha}{3h/4} = \frac{4}{3} \tan \alpha.$$

Length of string is $\dfrac{h}{\cos \beta} = h \sec \beta = \dfrac{h \sqrt{(16 \tan^2 \alpha + 9)}}{3}$.

Or:

Taking moments about B, for equilibrium,

$$Th \sec \alpha \sin(\beta - \alpha) = W \frac{h}{4}.$$

Resolving vertically, $\quad W = T \sin \beta.$

$$\therefore \sec \alpha \sin (\beta - \alpha) = \frac{\sin \beta}{4};$$

i.e., $4 \sec \alpha \; (\sin \beta \cos \alpha - \cos \beta \sin \alpha) = \sin \beta,$

$$4 \; (\sin \beta - \cos \beta \tan \alpha) = \sin \beta,$$

$$3 \sin \beta = 4 \cos \beta \tan \alpha,$$

$$\tan \beta = \tfrac{4}{3} \tan \alpha.$$

Complete as before.

5.6 A solid is formed by joining the square base, side h, of a pyramid height $h/2$, to the top face of a cube of edge h, as shown on the diagram. Show that the centre of mass of the solid is a distance $\frac{33}{56} h$ from the base of the cube. The solid stands with its base on a rough plane, which is gradually inclined to the horizontal, about an axis parallel to one edge of the cube. If the solid topples before slipping occurs, find the condition which must be satisfied by the coefficient of friction.

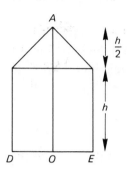

● Volume of cube = h^3. Volume of pyramid = $\dfrac{1}{3} (h^2) \left(\dfrac{h}{2}\right) = \dfrac{h^3}{6}$.

Ratio of masses is $6 : 1$.
Take all measurements from O along the line of symmetry.

Mass of cube is $6M$ acting at a distance $\dfrac{h}{2}$ from O.

Mass of pyramid is M acting at a distance $h + \dfrac{1}{4}\left(\dfrac{h}{2}\right) = \dfrac{9h}{8}$ from O.

Centre of mass of the solid is distance $\overline{x}$ from O.

Taking moments about O, $\quad 6M \left(\dfrac{h}{2}\right) + M \left(\dfrac{9h}{8}\right) = 7M\overline{x},$

$$\overline{x} = \tfrac{33}{56} h.$$

When the solid begins to topple, the centre of mass is vertically above D, and the normal reaction R acts through D.

$GO = \tfrac{33}{56} h, \quad DO = \tfrac{1}{2} h$

$$\tan \alpha = \dfrac{DO}{GO} = \dfrac{28}{33}.$$

Resolving normally to the plane, $\quad R = 7Mg \cos \alpha.$
Resolving parallel to the plane, $\quad F = 7Mg \sin \alpha.$

$$\dfrac{F}{R} = \tan \alpha.$$

Since the solid topples before it slides, $\quad \dfrac{F}{R} < \mu$ (coefficient of friction), $\quad \mu > \tfrac{28}{33}.$

5.7 Prove that the centre of mass of a uniform solid hemisphere whose radius is r lies on the radius of symmetry distant $\tfrac{3}{8} r$ from the centre of the plane face.

The hemisphere, whose weight is W, is placed with its circular face in contact with a smooth plane inclined at angle α to the horizontal. Equilibrium is maintained by a force P tangential to the curved surface and in the vertical pane containing the centre of mass and a line of greatest slope of the plane. The direction of P makes an angle β with the vertical.
(a) Find P in terms of W, α and β.
(b) If the magnitude of the reaction of the plane on the hemisphere is R, show that $R = W \sin \beta \operatorname{cosec} (\alpha + \beta)$.

(c) Find the distance of the line of action of R from the centre of the circular face of the hemisphere. (SUJB)

- Consider a disc distance x from the plane face.

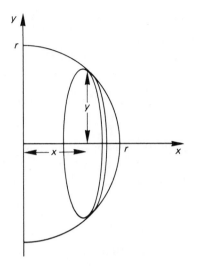

Mass $= \pi y^2 \rho \, \delta x$, where $x^2 + y^2 = r^2$.
Centre of mass $= (x, 0)$.
Mass of hemisphere $= \frac{2}{3} \pi r^3 \rho$.

Taking moments about the plane face, $\dfrac{2}{3} \pi r^3 \rho \bar{x} = \pi \rho \displaystyle\int_0^r (r^2 - x^2) x \, dx$,

i.e. $\dfrac{2}{3} r^3 \bar{x} = \left[\dfrac{r^2 x^2}{2} - \dfrac{x^4}{4} \right]_0^r$, $\dfrac{2}{3} r^3 \bar{x} = \dfrac{r^4}{4}$, $\bar{x} = \dfrac{3r}{8}$;

The centre of mass of a solid hemisphere is on the radius of symmetry, distant $\frac{3}{8} r$ from the plane face.

Force P makes an angle $[90 - (\alpha + \beta)]$ with the inclined plane.
(a) In equilibrium, resolving parallel to the plane,

$$ P \sin(\alpha + \beta) = W \sin \alpha, \quad \text{i.e.} \quad P = \frac{W \sin \alpha}{\sin(\alpha + \beta)}. $$

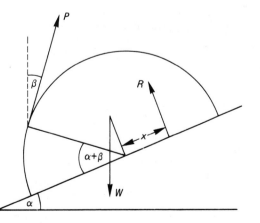

(b) Resolving perpendicular to the force P,

$$ W \sin \beta = R \sin(\alpha + \beta), \quad \text{i.e.} \quad R = W \sin \beta \operatorname{cosec}(\alpha + \beta). $$

(c) Let R act at a distance x from the centre of the circular face. Taking moments about the centre of the face,

$$W\left(\frac{3}{8}r\sin\alpha\right) = Pr - Rx,$$

i.e. $\dfrac{3}{8}Wr\sin\alpha = \dfrac{Wr\sin\alpha}{\sin(\alpha+\beta)} - \dfrac{Wx\sin\beta}{\sin(\alpha+\beta)},$

i.e. $x = \dfrac{r\sin\alpha}{8\sin\beta}[8 - 3\sin(\alpha+\beta)]$

$ = \dfrac{r}{8}\sin\alpha\,\mathrm{cosec}\,\beta\,[8 - 3\sin(\alpha+\beta)].$

5.8   A body is formed from a solid hemisphere of radius a and a solid right-circular cone of height b and base radius a, the hemisphere and cone having the same density and being joined with their plane faces completely in contact.

Show that the centre of gravity G of the body is at a distance $\dfrac{3b^2 + 8ab + 3a^2}{4(2a+b)}$

from the vertex of the cone. (Standard results on centres of gravity may be quoted.) Show that the body can always rest in equilibrium with the cone in contact with a horizontal plane.

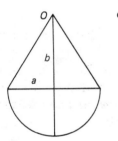

● Volume of cone $= \frac{1}{3}\pi a^2 b$.
  Volume of hemisphere $= \frac{2}{3}\pi a^3$.
  Ratio of volumes $= b : 2a$.
  Ratio of weights $= b : 2a$ since they have the same density.
  Let the weight of the cone be kb acting at a distance $\frac{3}{4}b$ from O.
  Let the weight of the hemisphere be $2ka$ acting at a distance $b + \frac{3}{8}a$ from O.
  Let the weight of the body, $k(2a+b)$, act at a distance $\bar{y}$ from O.
  Taking moments about O,

$$k(2a+b)\,\bar{y} = kb\left(\frac{3}{4}b\right) + 2ka\left(b + \frac{3}{8}a\right)$$

$$= \frac{k}{4}(3b^2 + 8ab + 3a^2),$$

$$\bar{y} = \frac{3b^2 + 8ab + 3a^2}{4(2a+b)}.$$

System will be in equilibrium if $OG\cos\alpha \leqslant OA$.
where α is the semi-vertical angle of the cone;

i.e. $OG\left(\dfrac{b}{OA}\right) \leqslant OA,\quad b(OG) \leqslant OA^2$

i.e. $\dfrac{b(3b^2 + 8ab + 3a^2)}{4(2a+b)} \leqslant OA^2 = a^2 + b^2,$

$3b^3 + 8ab^2 + 3a^2 b \leqslant 4(a^2 + b^2)(2a+b)$

$\leqslant 8a^3 + 8ab^2 + 4a^2 b + 4b^3,$

i.e. $b^3 + a^2 b + 8a^3 \geqslant 0,$ which is always true (sum of positive quantities).

72

5.3 Exercises

5.1 A uniform wire of length 12 cm and mass $12m$ grams is bent into a triangle ABC with $AB = 4$ cm, $BC = 5$ cm, $CA = 3$ cm. Find the distances of the centre of gravity G of the triangle from AB and AC.

The triangle hangs freely from A, and two masses m_1 grams and m_2 grams are attached to B and C respectively. In equilibrium, BC is horizontal. Show that $16m_1 - 9m_2 + 36m = 0$.

5.2 Show, without use of any result quoted in the booklet of formulae, that the centre of gravity of a uniform triangular lamina lies at the point of intersection of the medians.

A lamina $ABCD$, in the form of a rhombus, consists of two uniform triangular laminae ABC and ADC, joined along AC. Both triangles are equilateral of side $2a$. ABC is of mass $2m$ and ADC is of mass m. Determine the perpendicular distance of the centre of gravity G of the rhombic lamina from AC and from AB.

The rhombic lamina is suspended by a string attached to A. Determine:
(a) the tangent of the angle between AC and the vertical.
(b) the tension in the string when a mass is attached at D so that AC is vertical.

(AEB 1984)

5.3 Prove that the centre of mass of a uniform semicircular lamina of radius r is at a distance $4r/3\pi$ from the centre of the semicircle.

From a uniform lamina in the shape of a semicircle of radius $2a$, density ρ, a concentric semicircular portion of radius a is removed, and replaced by a semicircle radius a, density ρ_1. Find the ratio $\rho_1 : \rho$ if the centre of mass of the composite body is a distance $2a/\pi$ from the centre of the semicircles.

5.4 Prove, by integration, that the distance of the centre of mass of a uniform, solid, right, circular cone, of height h, from its plane base is $h/4$.

The cone is freely hinged at its vertex and is kept in equilibrium by a light rigid rod of length h joining the centre of the base to a point $h\sqrt{3}$ directly above the vertex. Show that the tension in the rod is $\dfrac{W\sqrt{3}}{4}$, where W is the weight of the cone.

Find the magnitude of the reaction at the hinge. (L)

5.5 Show, by integration, that the centre of mass of a uniform, solid, right circular cone of height h is at a distance $\dfrac{3h}{4}$ from the vertex.

Two uniform, solid, right circular cones, each with base radius a, have heights h and $3h$ and densities 3ρ and ρ. These cones are joined together with their bases coinciding to form a spindle. Show that the centre of mass is distance $\dfrac{11h}{4}$ from the vertex of the larger cone. Find a condition on a and h, if the spindle can rest in equilibrium with either curved surface in contact with a smooth horizontal table.

5.6 The figure represents a solid formed by the removal of a sphere of radius $a/2$ from a uniform solid hemisphere of base radius a. The point O denotes the centre of the plane base of the hemisphere, and C is that point of the hemisphere which is farthest from the base. Show that the centre of gravity of the solid is on OC at a distance $a/3$ from O.

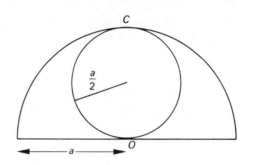

The solid is placed with its plane face on a rough plane inclined at an angle α to the horizontal. The coefficient of friction between the solid and the plane is $\frac{2}{3}$. Find the condition satisfied by $\tan\alpha$ in order that the solid does not slip immediately.

A gradually increasing force P is then applied to the solid. The force acts at the point C on the solid in a direction parallel to the line of greatest slope and up the plane. Given that the solid is of weight W, show that, when it is just about to topple:

$$P = W\cos\alpha + \tfrac{1}{3}W\sin\alpha.$$

Obtain the condition that has to be satisfied by $\tan\alpha$ in order that the body topples before it slips. (AEB 1983)

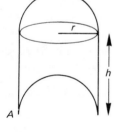

5.7 Prove that the centroid of a uniform hemisphere of radius a is at a distance $3a/8$ from O, the centre of the plane face of the hemisphere.

A stacking toy block is constructed from a solid, right, circular plastic cylinder of radius r and height h ($h > r$) which has a hemisphere of radius r removed from one end and added to the other (as shown in the diagram). Show that the centre of mass G is at a distance $\bar{x}$ from the centre of the hollow end, where $\bar{x} = \dfrac{3h + 4r}{6}$.

The block is suspended from a point A on the circumference of the hollowed end. Show that the axis of the block makes an angle θ with the vertical, where

$$\tan\theta = \frac{6r}{3h + 4r}.$$

5.4 Brief Solutions to Exercises

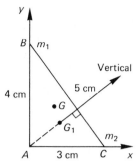

5.1 The triangle is right-angled at A. Considering axes along AC and AB,
AC has weight $3k$ at $(\frac{3}{2}, 0)$,
AB has weight $4k$ at $(0, 2)$,
BC has weight $5k$ at $(\frac{3}{2}, 2)$,　giving $\bar{x} = 1$,　$\bar{y} = 1.5$.
$12m$ at $(1, 1.5)$, m_1 at $(0, 4)$ and m_2 at $(3, 0)$

gives G_1 at $\left(\dfrac{3m_2 + 12m}{12m + m_1 + m_2},\ \dfrac{4m_1 + 18m}{12m + m_1 + m_2} \right)$; $(\bar{x}_1, \bar{y}_1)$.

If BC is horizontal and AG_1 vertical, $\angle G_1 AC = \angle ABC = \arctan \frac{3}{4}$.

But $\tan G_1 \hat{A} C = \dfrac{\bar{y}_1}{\bar{x}_1}$,　giving　$16m_1 - 9m_2 + 36m = 0$.

5.2　C.G. of element lies at midpoint of XY, i.e. on median from A to BC. This is true for all elements, so c.g. of triangle lies on this median. By symmetry, c.g. lies on all three medians, and is therefore at their intersection.

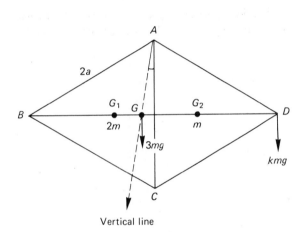

Vertical line

C.G. (G_1) of ABC is $\dfrac{1}{3}(2a \cos 30) = \dfrac{\sqrt{3}}{3} a$ from AC.

C.G. (G_2) of ADC is also $\dfrac{\sqrt{3}}{3} a$ from AC.

C.G. of mass $2m$ at G_1 and mass m at G_2 is at G, on $G_1 G_2$, distance $\dfrac{\sqrt{3}}{9} a$

from AC.　$BG = \dfrac{8\sqrt{3}}{9} a$.

Perp. dist. from AB to G is $\dfrac{8\sqrt{3}}{9} a \sin 30 = \dfrac{4\sqrt{3}}{9} a$.

(a) When suspended from A, AG is vertical, so $\tan C\hat{A}G = \dfrac{\sqrt{3}}{9}$.

(b) Attach mass km at D to make AC vertical.
　　Take moments about the midpoint of AC to get $k = \frac{1}{3}$.
　　Tension $= 3mg + kmg = \dfrac{10}{3} mg$.

5.3 See Worked Example 5.3 for proof with $\theta = \dfrac{\pi}{2}$.

Let mass of large semicircle be $4k\rho$.
Let mass of removed semicircle be $k\rho$.

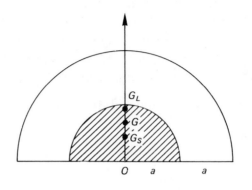

$$\left(4k\rho \text{ at } \frac{8a}{3\pi}\right) - \left(k\rho \text{ at } \frac{4a}{3\pi}\right) + \left(k\rho_1 \text{ at } \frac{4a}{3\pi}\right) \quad \text{gives} \quad k(3\rho + \rho_1) \text{ at } \frac{2a}{\pi}.$$

Taking moments about O gives $\quad 2\rho_1 = 10\rho, \quad \rho_1 : \rho = 5 : 1$.

5.4 See Worked Example 5.4 for the proof.
Cone is hinged at V. Moments about V give $\quad T = (W\sqrt{3})/4$.

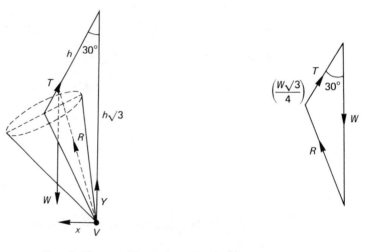

If the hinge reaction is R, consider triangle of forces:

$$R^2 = W^2 + \left(\frac{W\sqrt{3}}{4}\right)^2 - 2(W)\left(\frac{W\sqrt{3}}{4}\right)\cos 30, \qquad R = \frac{W\sqrt{7}}{4}.$$

5.5 For the proof see Worked Example 5.4.
Masses of cones are equal.

Taking moments about O gives $\overline{x} = \dfrac{11h}{4}$ from O.

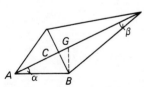

Let the semi-vertical angles of small and large cones be α and β.
When small cone rests on surface, equilibrium is possible if $AG \cos \alpha \leqslant AB$;
i.e. if $\frac{5}{4}h^2 \leqslant (AB)^2 = a^2 + h^2$,
i.e. if $h \leqslant 2a$.
When large cone rests on surface, spindle is always in equilibrium, since c.g. is within large cone. This implies that spindle can rest in equilibrium on either surface provided $h \leqslant 2a$.

5.6 Ratio of masses of sphere and hemisphere is 1 : 4. Let masses be M and $4M$. If centre of mass is distance $\bar{x}$ from O, taking moments about base gives

$$3M\bar{x} = 4M\left(\frac{3}{8}a\right) - M\frac{a}{2}, \quad \text{i.e.} \quad \bar{x} = \frac{a}{3}.$$

Resolving perpendicular and parallel to slope:

$$R = W\cos\alpha, \quad F = W\sin\alpha.$$

If solid not to slip immediately, $\dfrac{F}{R} \leqslant \mu = \dfrac{2}{3}$, i.e. $\tan\alpha \leqslant \dfrac{2}{3}$.

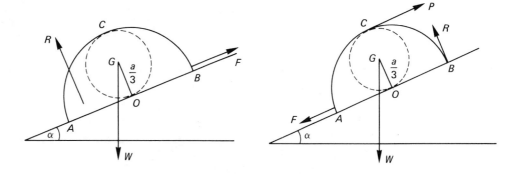

P applied: when solid is about to topple, the normal reaction will act through B.

Taking moments about B gives $P = W\cos\alpha + \dfrac{1}{3}W\sin\alpha$.

Resolving parallel to and perpendicular to the slope gives

$$\frac{F}{R} = 1 - \frac{2}{3}\tan\alpha < \frac{2}{3} \qquad \text{for no slipping,}$$

$$\tan\alpha > \frac{1}{2}.$$

5.7 For proof see Worked Example 5.7.
Volume of cylinder = $\pi r^2 h$.
Volume of hemisphere = $\frac{2}{3}\pi r^3$.
Ratio of masses is $3h : 2r$.
Mass of cylinder is $3hk$, mass of hemisphere is $2rk$.

$$\left(3hk \text{ at } \frac{h}{2}\right) - \left(2rk \text{ at } \frac{3}{8}r\right) + \left(2rk \text{ at } h + \frac{3}{8}r\right) \quad \text{gives } 3hk \text{ at } \frac{3h + 4r}{6} \quad (= \bar{x}).$$

The angle made with the vertical is θ where $\tan\theta = \dfrac{r}{\bar{x}} = \dfrac{6r}{3h + 4r}$.

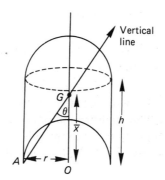

6 Constant Acceleration. Velocity—Time Graphs. Power

Kinematics of a particle moving in a straight line including graphical treatment.
Derivation and use of constant acceleration formulae.
Newton's laws of motion.
Motion of a particle under the action of a constant force.
Power.

6.1 Fact Sheet

(a) Constant-acceleration Formulae

If a point is moving in a straight line and the distance of the point from O at time t is x (or s) then

$$v = \frac{dx}{dt}, \qquad a = \frac{dv}{dt} = \frac{d^2x}{dt^2} = v\frac{dv}{dx}.$$

With constant acceleration a,

$$v = \int a\,dt = at + u, \qquad\qquad v = u + at; \qquad (1)$$

$$x = \int v\,dt = ut + \tfrac{1}{2}at^2 + c, \qquad s \text{ or } x = ut + \tfrac{1}{2}at^2. \qquad (2)$$

From (1) and (2), $\quad s = \dfrac{u+v}{2}\,t \quad$ and $\quad v^2 = u^2 + 2as \quad$ can be obtained.

(b) Newton's Laws

(i) Every body will remain at rest or continue to move with constant velocity unless an external force is applied to it.

or If a body has an acceleration there is a force acting to cause it.

At constant, or maximum speed, the sum of forces = 0.

(ii) When an external force is applied to a body of constant mass, the force produces an acceleration directly proportional to the force.

or The sum of forces in the direction of motion = mass × acceleration in that direction ($F = ma$).

(iii) To every action there is an equal and opposite reaction.

(c) Power

Power is the rate of working, measured in watts.
For a given speed of v metres per second and tractive force of F newtons, the power developed = Fv watts.

(d) Velocity–Time Graphs

If the units of time are the same in velocity and time then:
(i) The gradients of the lines represent accelerations.
(ii) The area under the graph represents the distance travelled.
   It is not necessary to use SI units for these graphs.

6.2 Worked Examples

6.1 The brakes of a train, which is travelling at 108 km h^{-1}, are applied as the train passes point A. The brakes produce a constant retardation of magnitude $3f$ m s^{-2} until the speed of the train is reduced to 36 km h^{-1}. The train travels at this speed for a distance and is then uniformly accelerated at f m s^{-2} until it again reaches a speed of 108 km h^{-1} as it passes point B. The time taken by the train in travelling from A to B, a distance of 4 km, is 4 minutes.

Sketch the speed–time graph for this motion and hence calculate

(a) the value of f;

(b) the distance travelled at 36 km h^{-1}. (L)

● 108 km h^{-1} = 30 m s^{-1} 36 km h^{-1} = 10 m s^{-1}.

If train takes t_1 seconds to decelerate then it takes $3t_1$ seconds to accelerate.
   Therefore it is at constant speed 10 m s^{-1} for $240 - 4t_1$ seconds.

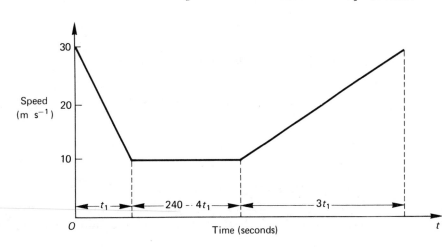

Distance travelled = area under graph
$$= 20t_1 + 10(240 - 4t_1) + 60t_1 = 4000 \text{ m},$$
$$\Rightarrow \quad 40t_1 = 1600, \quad t_1 = 40.$$
Therefore train takes 40 seconds to decelerate from 30 m s^{-1} to 10 m s^{-1}, a deceleration of 0.5 m s^{-2},

$$\text{hence } f = \tfrac{1}{6} \text{ m s}^{-2}.$$

Distance travelled at 36 km h^{-1} = 10(240 - 160) = 800 m.

6.2 A car moves from rest with constant acceleration $3f$ from W to X, then continues from X to Y with acceleration f, from Y to Z with constant speed V. If the times taken for WX, XY and YZ are each equal to T, find
(a) T in terms of V and f,
(b) the ratio of the distance $WX : XY : YZ$,
(c) the retardation if the car comes to rest in a further time T,
(d) the total distance travelled in terms of V and f.

● *Velocity–time graph*

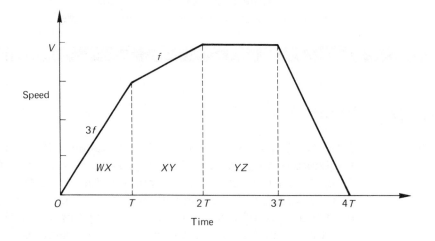

● (a) Speed acquired in time T with acceleration $3f$ is $3fT$.
In further time T, with acceleration f,
speed $V = 3fT + fT = 4fT$.
Constant speed $V = 4fT \quad \Rightarrow \quad T = V/4f$.
(b) Distances are given by areas under the graph
Distance $WX = (\tfrac{1}{2}T)(\tfrac{3}{4}V) = \tfrac{3}{8}TV$.
Distance $XY = \tfrac{1}{2}T(\tfrac{3}{4}V + V) = \tfrac{7}{8}TV$.
Distance $YZ = TV$.
Therefore $WX : XY : YZ = \tfrac{3}{8} : \tfrac{7}{8} : 1$ or $3 : 7 : 8$.
(c) To slow down from V to rest in time T seconds, retardation $= V/T = 4f$.

(d) Distance travelled while retarding $= \dfrac{1}{2}VT$.

$$\text{Therefore total distance travelled} = \frac{3}{8}TV + \frac{7}{8}TV + TV + \frac{1}{2}TV = \frac{11}{4}TV$$

$$= \frac{11V^2}{16f}.$$

6.3 On a recent journey from the foundry to the nearest port a lorry with a heavy load accelerated uniformly during the first 4 minutes until it reached a speed of 12 km h^{-1}. Because of traffic conditions it had to maintain this speed for the next 20 minutes until it reached the motorway where it accelerated uniformly, reaching its maximum speed of 20 km h^{-1} after a further 7.5 minutes. The lorry maintained this speed until it applied its brakes 2 km before the end of the motorway to give a uniform retardation. It came to rest at the end of the motorway having travelled 49 km on the motorway.

Draw a velocity–time graph and calculate, either from the graph or using the equations of motion,

(a) the rates of acceleration and retardation,
(b) the total distance travelled,
(c) the average speed.

● *Velocity–time graph*

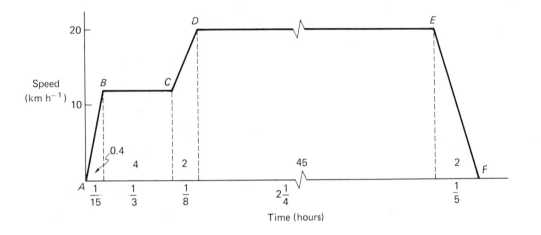

(a) Using $a = \dfrac{v - u}{t}$,

   first acceleration (A to B) $= \dfrac{12}{1/15} = 180$ km h^{-2} (0.0139 m s^{-2});

   second acceleration (C to D) $= \dfrac{8}{1/8} = 64$ km h^{-2} (0.0049 m s^{-2}).

   Using $v^2 = u^2 + 2as$ $\Rightarrow$ $a = \dfrac{v^2 - u^2}{2s}$,

   retardation (E to F) $= \dfrac{(20)^2}{(2)(2)} = 100$ km h^{-2} (0.0077 m s^{-2}).

(b) Distance travelled from A to $B = \frac{1}{2}(12)(\frac{1}{15})$ km $= 400$ m.
    Distance travelled from B to $C = 12(\frac{1}{3}) = 4$ km.
    Distance travelled on the motorway C to $F = 49$ km.
    Thus total distance travelled $= 53.4$ km.

(c) Distance travelled from C to $D = \dfrac{(12 + 20)}{2}\dfrac{1}{8}$

    $= 2$ km.

   So distance travelled at constant speed on the motorway $= 45$ km.

   Time taken from D to $E = \dfrac{45}{20} = 2.25$ h.

   Time taken from E to $F = \dfrac{20}{100} = \dfrac{1}{5}$ h $\left(\text{using } t = \dfrac{v - u}{f}\right).$

81

Total time taken = $\frac{1}{15} + \frac{1}{3} + \frac{1}{8} + \frac{9}{4} + \frac{1}{5}$ = 2.975 h.

Therefore the average speed = $\dfrac{\text{distance}}{\text{time}}$ = $\dfrac{53.4}{2.975}$ = 17.9 km h^{-1}.

6.4 A car of mass 1.2 tonnes is travelling up a slope of 1 in 150 at a constant speed of 10 m s^{-1}. If the frictional and air resistances are 100 N, calculate the power exerted by the engine. The car descends the same slope working at a rate of 2 kW. What will be its acceleration when its speed is 20 m s^{-1} if the resistances are the same? If the engine is shut off when the speed of the car is 25 m s^{-1} as it descends the slope, how long will it be before the car comes to rest?   (SUJB)

- Take g = 10 m s^{-2}.
  Ascending:

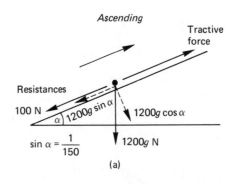

(a)

Accelerating force = tractive force − 100 − 1200g sin α
= 0 when speed is constant.
Therefore tractive force = 100 + 80 = 180 N.
Power of the engine = tractive force × speed
= 180 × 10
= 1.8 kW.

Descending:
If power = 2 kW and speed = 20 m s^{-1} then tractive force = 100 N.
Accelerating force = 1200g sin α + 100 − 100 = 80 N.

Therefore acceleration = $\dfrac{80}{1200}$ m s^{-2} = $\dfrac{1}{15}$ m s^{-2}.

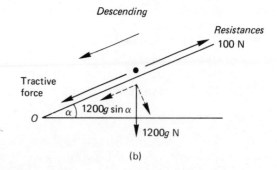

(b)

When the engine is shut off the accelerating force is 80 − 100 = −20 N.

Therefore the car has a retardation of $\dfrac{20}{1200}$ m s^{-2} = $\dfrac{1}{60}$ m s^{-2}.

Therefore the car will take $\dfrac{v}{a}$ s = 60 × 25 s to come to rest, i.e. 25 minutes.

82

6.5 A car of mass 1500 kg is travelling along a straight horizontal road at a constant speed of 40 m s^{-1} against a resistance of 900 N. Calculate, in kW, the power being exerted.

Calculate the power required to go down a hill of 1 in 25 (along the slope) at a steady speed of 40 m s^{-1} against the same resistance.

Given that the resistance is proportional to the square of the speed of the car, calculate the acceleration of the car up this hill with the engine working at 18 kW at the instant when the speed is 20 m s^{-1}.

● Take $g = 10$ m s^{-2}.
At constant speed the accelerating force = 0.
Therefore the tractive force = resistances.
∴ On a level road tractive force = 900 N.
At 40 m s^{-1} the power being exerted = $Fv = 900 \times 40$ W
$$= 36 \text{ kW}.$$
To go down the hill at a steady speed,
tractive force + $mg \sin \alpha = 900$;
$mg \sin \alpha = 15\,000/25 = 600$.
Therefore the tractive force = 300 N.
The power exerted at 40 m s^{-1} = $Fv = 300 \times 40$ W = 12 kW.
If resistance is proportional to the square of the speed then $R = kv^2$.
When $R = 900$, $v = 40$, so $k = \frac{9}{16}$ (= 0.5625).

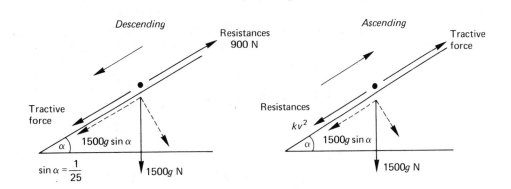

To go up the hill,
When $v = 20$ m s^{-1} and power = 18 kW, tractive force is $\dfrac{18 \times 10^3}{20} = 900$ N.

Resistance = $0.5625 (20)^2 = 225$ N.
Tractive force − ($mg \sin \alpha$ + resistance) = accelerating force,
⇒ $900 - (600 + 225) = 1500a$.

Acceleration $a = \dfrac{75}{1500}$ m s^{-2} = 0.05 m s^{-2}.

6.6 A bus of mass 18 tonnes travels up a slope of gradient $\sin^{-1} \frac{1}{50}$ against a resistance of 0.1 N per kilogram. Find the tractive force required to produce an acceleration of 0.05 m s^{-2} and the power which is developed when the speed is 10 m s^{-1}.

A second bus of mass 25 tonnes experiencing the same resistance and with a maximum power 120 kW follows the first bus up the slope. If the first bus maintains the same power while on the slope, will the distance between the buses decrease when both are travelling at maximum speed?

- Take $g = 10$ m s^{-2}.

First bus:

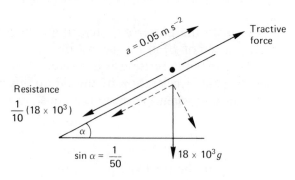

Mass = 18×10^3 kg, resistance $R = 1.8 \times 10^3$ N.

Component of weight down the slope $W = \dfrac{180 \times 10^3}{50} = 3.6$ kN.

Accelerating force = $18 \times 10^3 \times 0.05$ N = 0.9 kN.

Accelerating force = tractive force − (R + component of weight),

$\qquad 0.9 \times 10^3$ = tractive force − $(1.8 \times 10^3 + 3.6 \times 10^3)$.

Hence

$\qquad$ tractive force = $(0.9 + 1.8 + 3.6) \times 10^3$ N

$\qquad\qquad = \underline{6.3 \text{ kN.}}$

When the speed is 10 m s^{-1}, power = $10 \times 6.3 \times 10^3$ W

$\qquad\qquad = \underline{63 \text{ kW.}}$

At maximum speed, tractive force = R + component of weight

$\qquad\qquad = 1.8$ kN + 3.6 kN = 5.4 kN.

Power = 63 kW,

$\Rightarrow \quad$ maximum speed = $\dfrac{\text{power}}{\text{tractive force}} = \dfrac{63}{5.4} = \tfrac{35}{3}$ m s^{-1}

$\qquad\qquad\qquad = 11\tfrac{2}{3}$ m s^{-1}.

Second bus:

Mass = 25×10^3 kg, resistance $R = 2.5 \times 10^3$ N.

Component of weight down the slope $W_s = \dfrac{25 \times 10^4}{50}$ N = 5 kN.

At maximum speed accelerating force = 0,

$\qquad\qquad$ tractive force = $R + W_s$

$\qquad\qquad\qquad = (2.5 + 5)$ kN = 7.5 kN.

But power = tractive force × speed;

$\quad 120 \times 10^3 = 7.5 \times 10^3 \times$ speed.

Therefore the maximum speed of the bus is 16 m s^{-1}.

Since the second bus travels faster than the first the distance between them will decrease.

6.7 Prove, using definitions, algebra and/or calculus, that

$$v = u + at \qquad \text{and} \qquad s = ut + \tfrac{1}{2} at^2$$

for motion in a straight line with uniform acceleration. A car A, moving with uniform velocity u_A along a straight road, passes at a point X a car B, moving in the same direction with velocity u_B and uniform acceleration b. B overtakes A at point Y.

(a) Prove that the time taken to reach Y is

$$\frac{2(u_A - u_B)}{b}.$$

(b) Find the distance XY.

(c) After passing X but before reaching Y, find the distance between the cars at time T, and prove that the maximum distance between them is $\dfrac{(u_A - u_B)^2}{2b}$.

At Y, A now accelerates with uniform acceleration a and B changes to uniform velocity, keeping the velocity it had on arrival at Y. If A overtakes B again after it has travelled a further distance equal to XY, prove that $a = \dfrac{(2u_A - u_B)\,b}{u_A}$.

(SUJB)

- $\dfrac{dv}{dt} = a$, so, integrating, $v = at + C$.

  $v = u$ when $t = 0$ so $C = u$ and $v = u + at$.
  Integrating again, $s = ut + \frac{1}{2}at^2 + D$.
  If $s = 0$ when $t = 0$, $D = 0$ so $s = ut + \frac{1}{2}at^2$.

  Let $t = 0$ when A passes B at X.
  Distances travelled by A and B in time t are $s_A = u_A t$, $s_B = u_B t + \frac{1}{2}bt^2$.
  When B overtakes A at Y after a time t_1,

  $$s_A = s_B, \quad u_A t_1 = u_B t_1 + \tfrac{1}{2}bt_1^2 \quad \Rightarrow \quad 2(u_A - u_B)t_1 = bt_1^2.$$

(a) $t_1 = 0$ (when A passes B at X),

   or $t_1 = \dfrac{2(u_A - u_B)}{b}$ (B passes A at Y).

(b) Distance $XY = u_A t_1 = \dfrac{2u_A(u_A - u_B)}{b}$.

(c) Distance between cars at time $T = s_A - s_B$
   $$= (u_A - u_B)T - \tfrac{1}{2}bT^2.$$
   Maximum distance between cars occurs when $v_A = v_B$,

   then $u_A = u_B + bT$, $\Rightarrow$ $T = \dfrac{(u_A - u_B)}{b}$

   and maximum distance between the cars is

   $$\frac{(u_A - u_B)^2}{b} - \frac{b}{2}\frac{(u_A - u_B)^2}{b^2} = \frac{1}{2b}(u_A - u_B)^2.$$

   At Y, A has speed u_A, acceleration a,
   B has speed $u_B + bt_1 = u_B + 2(u_A - u_B)$
   $$= 2u_A - u_B = u_{B_1}.$$
   From (b), the distance travelled before A overtakes B is

   $$\frac{2u_{B_1}(u_{B_1} - u_A)}{a} = XY \quad \text{(given)}.$$

   Thus $\dfrac{2(2u_A - u_B)(u_A - u_B)}{a} = \dfrac{2u_A(u_A - u_B)}{b}$.

   Therefore $a = \dfrac{b(2u_A - u_B)}{u_A}$.

6.3 Exercises

6.1 (multiple choice) Points P_1 and P_2 start together at the origin and move along the x-axis so that their respective displacements x_1 and x_2 in metres at time t seconds are given by

$$x_1 = t, \quad x_2 = t^2.$$

1 P_2 begins to move ahead of P_1.
2 P_1 and P_2 will coincide again at some later time.
3 P_1 and P_2 have equal velocities at time $t = 0.5$.

A, 1, 2, 3 are correct;   B, only 1 and 2 are correct,   C, only 2 and 3 are correct.
D, only 1 is correct;   E, only 3 is correct.

(L)

6.2 Given that s, defined by $s = ut + \frac{1}{2}at^2$, where u and a are constants, represents the displacement of a particle at time t show, by differentiation, that u is the velocity at time $t = 0$ and that the acceleration is equal to a.

A train starting from rest is uniformly accelerated during the first minute of its journey when it covers 600 m. It then runs at a constant speed until it is brought to rest in a distance of 1 km by applying a constant retardation.
(a) Find the maximum speed attained by the train.
(b) Determine the magnitude of the retardation.
(c) Given that the total journey time is 5 minutes determine the distance covered at constant speed.
(d) Given that the magnitude of the retardation, instead of being constant is directly proportional to the speed and the train comes to rest from the same constant speed in a distance of 500 m, find the magnitude of the retardation in m s^{-2} when the train's speed is 10 m s^{-1}.   (AEB 1982)

6.3 The diagram shows the velocity–time graph of a particle moving in a straight line. Along the t-axis the scale is in seconds, on the v-axis the scale is in m s^{-1}.

For $30 < t \leqslant 40$ the arc ABC has the equation

$$v^2 + (t - 30)^2 = 10^2.$$

Find the acceleration of the particle, in terms of t, during the intervals

$$0 < t \leqslant 5, \quad 5 < t \leqslant 30, \quad 30 < t \leqslant 40.$$

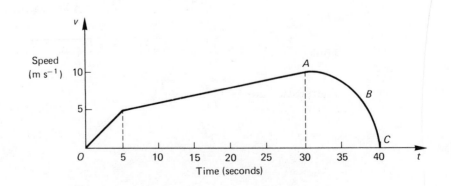

6.4 The mass of a car is 800 kg and the total resistance to its motion is constant and equal to a force of 320 N.
(a) Find, in kW, the rate of working of the engine of the car when it is moving along a level road at a constant speed of 25 m s^{-1}.
(b) What is the acceleration in m s^{-2}, when the car is moving along a level road at 20 m s^{-1} with the engine working at 11 kW?
(c) What is the maximum speed attained by the car with the engine working at this rate?

6.5 A car of mass 10^3 kg is travelling up a slope, inclination sin^{-1} (0.05), at a constant speed of 25 m s^{-1}. The engine is generating its maximum power of $\frac{5}{4} \times 10^5$ W. Show that, excluding gravity, the car is experiencing a resistance to motion of 4.5×10^3 N. (Let $g = 10$ m s^{-2}.)
  Assuming that the resistance is proportional to the speed of the car determine:
(a) the maximum speed of the car on the level road,
(b) the maximum acceleration of the car when it is travelling at 10 m s^{-1} on a level road.

6.6 A car has an engine capable of developing 15 kW. The maximum speed of the car on a level road is 120 km h^{-1}. Calculate the total resistance in newtons at this speed.
  Given that the mass of the car is 1000 kg and that the resistance to motion is proportional to the square of the speed, obtain the rate of working, in kW to two decimal places, of the engine when the car is moving at a constant speed of 40 km h^{-1} up a road of inclination θ, where $\sin \theta = 1/25$.　(L)
(Take g as 9.8 m s^{-2}.)

6.7 A car A moves with a constant acceleration f from rest up to its maximum speed u. Just as it starts it is overtaken by a car B moving with constant speed v. Given that A catches up with B in a distance $a \left(> \dfrac{u^2}{2f} \right)$, show that A has been travelling with speed u for a time $\dfrac{a}{v} - \dfrac{u}{f}$. Show also that $u^2 v - 2afu + 2afv = 0$.
  Hence find u in terms of a, f and v, explaining carefully how you decide which root to choose of the quadratic equation.　(L)

6.4 Brief Solutions to Exercises

6.1 $x_1 = t$, $x_2 = t^2$.
$x_1 = x_2$　when　$t = 0$ or 1.　(2 correct)
$\dot{x}_1 = 1$, $\dot{x}_2 = 2t$,
when $t = 0$, $\dot{x}_1 = 1$, $\dot{x}_2 = 0$,　(1 incorrect)
when $t = 0.5$, $\dot{x}_1 = \dot{x}_2 = 1$.　(3 correct)

Answer C

6.2 (a) $s = 600$, $u = 0$, $t = 60$ $\Rightarrow$ $v = 20$. Max. speed $= 20$ m s^{-1}.

(b) $u = 20$, $s = 1000$, $v = 0$ $\Rightarrow$ $a = 0.2$, retard. is 0.2 m s^{-2} $(v^2 = u^2 - 2as)$.
(c) $t = 100$ s for retardation $(v = u - at)$.

Time at constant speed = 300 − (100 + 60) = 140 s.
Distance at constant speed = 20 × 140 = 2800 m.

(d) Retardation = kv, $\Rightarrow$ $v\dfrac{dv}{ds} = -kv$

$$\Rightarrow \quad \dfrac{dv}{ds} = -k, \quad v = -ks + c.$$

Initially, $v = 20$, $s = 0$ $\Rightarrow$ $c = 20$.
Finally, $v = 0$, $s = 500$ $\Rightarrow$ $k = \frac{1}{25}$.
Thus retardation = $\dfrac{v}{25}$.
When $v = 10$ retardation = $\frac{2}{5}$ m s^{-2}.

6.3 (a) $0 < t \leqslant 5$, gradient = 1 $\Rightarrow$ acceleration = 1 m s^{-2}.
(b) $5 < t \leqslant 30$, gradient = $\frac{5}{25}$ $\Rightarrow$ acceleration = 0.2 m s^{-2}.

(c) $v^2 = 10^2 - (t - 30)^2 = 60t - t^2 - 800$, $\quad 2v\dfrac{dv}{dt} = -2(t - 30)$.

Therefore acceleration = $\dfrac{-(t - 30)}{\sqrt{(60t - t^2 - 800)}}$ m s^{-2}.

6.4 (a) Tractive force = resistance = 320 N.
Power = rate of work = Fv = (320) (25) = 8 kW.
(b) Tractive force = $\dfrac{11 \times 10^3}{20}$ = 550 N.

Acceleration = $\dfrac{\text{tractive force} - \text{resistance}}{\text{mass}}$ = 0.2875 m s^{-2}

(c) At maximum speed t.f. = resistance = 320 N.
Max. speed = $\dfrac{\text{power}}{\text{t.f.}}$ = 34.375 m s^{-1}.

6.5 Tractive force = $\dfrac{1.25 \times 10^5}{25}$ N = 5000 N.

Tractive force = resistance + component of weight
$\Rightarrow$ resistance = 5000 − 500 N = 4500 N.
(a) $R = kv = 180v$ from the first part.
On level road t.f. = resistance = $180v$.
Maximum speed = $\dfrac{\text{power}}{\text{t.f.}}$, i.e. $v = \dfrac{1.25 \times 10^5}{180v}$, $v = 26.4$ m s^{-1}.
(b) $v = 10$, resistance = 1800, t.f. = $\dfrac{1.25 \times 10^5}{10}$ = 12 500 N.

Accelerating force = 12 500 − 1800 = 10 700 N.
Maximum acceleration = $\dfrac{10\,700}{\text{mass}}$ = 10.7 m s^{-2}.

6.6   Speed = $\dfrac{100}{3}$ m s^{-1}. Tractive force = $\dfrac{\text{power}}{\text{speed}}$ = 450 N.

At max. speed, resistance = tractive force = 450 N.

$R = kv^2 \quad \Rightarrow \quad k = \dfrac{R}{v^2} \quad \Rightarrow \quad k = 0.405,$

$R = 0.405v^2.$

At 40 km h^{-1} ($= \frac{100}{9}$ m s^{-1}), $R = 50$ N.
Tractive force $= R + mg\sin\theta = 442$ N.

Therefore power $= 442 \times \dfrac{100}{9}$ W = 4.91 kW.

6.7 A takes time $\dfrac{u}{f}$ to reach maximum speed in distance $\dfrac{u^2}{2f}$.

Distance travelled by A in time $\left(\dfrac{u}{f} + t\right)$ is $\dfrac{u^2}{2f} + ut = a.$ $\qquad\qquad$ (1)

Distance travelled by B in time $\left(\dfrac{u}{f} + t\right)$ is $\dfrac{uv}{f} + vt = a.$ $\qquad\qquad$ (2)

From (2), $\quad \Rightarrow \quad t = \dfrac{a}{v} - \dfrac{u}{f}.$ $\qquad\qquad$ (3)

Substitute for t in (1) to get $\quad u^2v - 2afu + 2afv = 0,$

$$u = \frac{2af \pm \sqrt{(4a^2f^2 - 8afv^2)}}{2v} = \frac{af}{v} \pm \frac{\sqrt{(a^2f^2 - 2afv^2)}}{v}.$$

Since $t > 0$, using (3), $\quad \dfrac{af}{v} > u \quad$ so take $-$ve sign.

$$u = \frac{af}{v}\left[1 - \left(1 - \frac{2v^2}{af}\right)^{1/2}\right].$$

7 Projectiles

The equations of motion of a particle in vector form. Free motion under gravity. Motion in a parabola with constant acceleration.

7.1 Fact Sheet

(a) Vector Approach

Newton's second law $F = ma$ can be generalized to $\mathbf{F} = m\ddot{\mathbf{r}}$ for motion in two or three dimensions.
For motion under gravity,

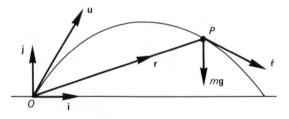

$$\mathbf{F} = m\ddot{\mathbf{r}} = m\mathbf{g}, \qquad \text{therefore} \qquad \ddot{\mathbf{r}} = \mathbf{g}.$$

Integrating, $\mathbf{v} = \dot{\mathbf{r}} = \mathbf{u} + t\mathbf{g}$, where $\mathbf{u}$ is the initial velocity.

Integrating, $\mathbf{r} = t\mathbf{u} + \frac{1}{2}t^2\mathbf{g} + \mathbf{r}_0$,

where $\mathbf{r}_0$ is the position vector at $t = 0$ (usually zero).

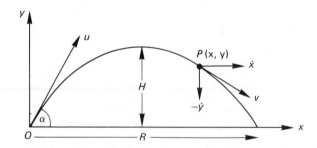

(b) Cartesian Approach

If $\mathbf{i}$ and $\mathbf{j}$ are unit vectors horizontally and vertically upwards respectively, then

$$\mathbf{r} = x\mathbf{i} + y\mathbf{j}$$

and $\ddot{\mathbf{r}} = \mathbf{g}$ may be written $\ddot{x} = 0$, $\ddot{y} = -g$.

If the particle is projected with speed u at an angle α to the horizontal, then

$$\dot{x} = u\cos\alpha \qquad (1), \qquad\qquad \dot{y} = u\sin\alpha - gt \qquad (2),$$

$$x = (u\cos\alpha)\,t \qquad (3), \qquad\qquad y = (u\sin\alpha)\,t - \tfrac{1}{2}gt^2 \qquad (4),$$

taking $x = 0$, $y = 0$ as the point of projection.

(c) Velocities

From (2)
$$
\begin{aligned}
\dot{y}^2 &= u^2\sin^2\alpha - 2ugt\sin\alpha + g^2t^2 \\
&= (u\sin\alpha)^2 - 2g\left(ut\sin\alpha - \tfrac{1}{2}gt^2\right) \\
\dot{y}^2 &= (u\sin\alpha)^2 - 2gy. \qquad\qquad (5) \\
\dot{x}^2 + \dot{y}^2 &= u^2\cos^2\alpha + u^2\sin^2\alpha - 2gy,
\end{aligned}
$$

or
$$v^2 = u^2 - 2gy, \qquad\qquad (6)$$

where v is the speed at any height y.

This can be rewritten as $\tfrac{1}{2}mu^2 - \tfrac{1}{2}mv^2 = mgy$, or

$$\text{loss of kinetic energy = gain in potential energy.}$$

This is the conservation of energy equation for a particle.

(d) Range R on the Horizontal Plane

The range R is the horizontal displacement (x) of the particle when its vertical displacement (y) is again zero.

From (4), $y = 0$ when $t = 0$ (initially), and when $t = \dfrac{2u\sin\alpha}{g}$.

This is the 'time of flight'.

Substituting in (3) gives

$$R = u\cos\alpha\left(\frac{2u\sin\alpha}{g}\right) = \frac{2u^2\sin\alpha\cos\alpha}{g}. \qquad\qquad (7)$$

This can be changed to $R = \dfrac{u^2\sin 2\alpha}{g}$ to find the two possible (complementary) angles of projection for a given range.

(e) Maximum Height H

Using equation (5), at maximum height, $\dot{y} = 0$,

$$u^2\sin^2\alpha = 2gH \qquad \Rightarrow \qquad H = \frac{u^2\sin^2\alpha}{2g}. \qquad\qquad (8)$$

There are several alternative methods for finding H and R.

(f) Equation of the Trajectory

The trajectory is the parabolic path followed by the particle.
Eliminate t from (3) and (4).

From (3)
$$t = \frac{x}{u\cos\alpha}.$$

In (4),
$$y = u \sin \alpha \left(\frac{x}{u \cos \alpha} \right) - \frac{gx^2}{2u^2 \cos^2 \alpha}$$

$$= x \tan \alpha - \frac{gx^2}{2u^2} \sec^2 \alpha$$

or
$$y = x \tan \alpha - \frac{gx^2}{2u^2} (1 + \tan^2 \alpha). \qquad (9)$$

This may be used in questions where a particle has to pass through, or strike, a given point.

Differentiating (9) with respect to x gives

$$\frac{dy}{dx} = \tan \alpha - \frac{gx}{u^2} (1 + \tan^2 \alpha).$$

This gives the direction of motion at a given horizontal displacement.

7.2 Worked Examples

7.1 A ball is projected with speed 40 m s^{-1} at an angle of 30° to the horizontal.
(a) Find the time taken for the ball
   (i) to travel 20 m horizontally,
   (ii) to reach a height of 20 m.
(b) Find the horizontal range of the ball.
   (Take $g = 10$ m s^{-2}.)

● Let **i** and **j** be unit vectors horizontally and vertically respectively.

$\ddot{\mathbf{r}} = -g\mathbf{j}$.

Integrating, $\dot{\mathbf{r}} = \mathbf{v} = \mathbf{u} - gt\mathbf{j}$.
Initial velocity $\mathbf{u} = 40 \cos 30\mathbf{i} + 40 \sin 30\mathbf{j}$
$\qquad\qquad\quad = 20\sqrt{(3)}\mathbf{i} + 20\mathbf{j}$.
So $\mathbf{v} = 20\sqrt{3}\mathbf{i} + (20 - gt)\mathbf{j}$,
$\quad \mathbf{r} = 20\sqrt{3}t\mathbf{i} + \left(20t - \frac{g}{2}t^2 \right) \mathbf{j}$.

In component form $x = 20\sqrt{3}t$, $y = \left(20t - \frac{g}{2}t^2 \right)$.

(a) (i) When $20\sqrt{3}t = 20$, $t = \frac{1}{\sqrt{3}} = 0.577$ s.

It takes 0.577 seconds to travel 20 m horizontally.
   (ii) When $20t - 5t^2 = 20$,
   $t^2 - 4t + 4 = 0$, $\Rightarrow$ $t = 2$ (twice).
   It takes 2 seconds to reach a height of 20 m.
   Since at this height the equation for t has equal roots, it must be the maximum height.

(b) $y = 0$ when $t = 0$ (initially) and when $t = \frac{40}{g} = 4$ s.

(Alternatively, total time = twice the time taken to reach the maximum height.)
When $t = 4$, $x = 80\sqrt{3} = 138.6$ m.
Horizontal range = 138.6 m.

7.2   A cricketer hits a ball so that it first lands on the ground at a point P at a distance 75 m from him. At the highest point of its path the ball reaches a height of 20 m. Assuming that the ball is projected from ground level determine

(a) the horizontal and vertical components of the velocity of projection;

(b) the tangent of the angle between the horizontal and the direction of motion of the ball 1 s after it has been hit;

(c) the furthest distance from P that a fielder who can run at 8 m s^{-1} can stand, in order that, starting when the ball is hit, he can arrive at P before the ball lands;

(d) the tangent of a different angle of projection which is such that the ball, when projected with the same initial speed, again first lands at P.
(Let $g = 10$ m s^{-2}).   (AEB 1982)

● (a) Let the initial components of velocity be v_x and v_y.

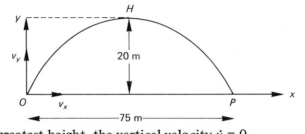

At the greatest height, the vertical velocity $\dot{y} = 0$.

Using $\dot{y}^2 = v_y^2 - 2gy$, $v_y^2 = 2g(20)$, $v_y = 20$ m s^{-1}.

Time taken to reach highest point = $\dfrac{v_y}{g}$ = 2 s.

Therefore time of flight = 4 s (symmetry).

$v_x = \dfrac{75}{4} = 18.75$ m s^{-1} (remains constant).

(b) After one second, $\dot{y} = 20 - g = 10$ m s^{-1}.

The tangent of the angle between the path and the horizontal after

1 second = $\dfrac{10}{18.75}$ or $\dfrac{8}{15}$.

(c) Since the time of flight is 4 s, the fielder can be up to 32 m from P.

(d) For any given range there are (usually) two angles of projection which are complementary.

For given projection $\tan \alpha = \dfrac{20}{18.75} = \dfrac{16}{15}$.

Second angle has a tangent of $\dfrac{15}{16}$.

7.3   A particle is projected with speed u at an angle of elevation α to the horizontal. Given that R is the range attained on a horizontal plane through the point of projection, and h is the maximum height of the trajectory, prove that

$$R = 2c \sin \alpha \cos \alpha, \qquad 2h = c \sin^2 \alpha, \qquad \text{where } c = u^2/g.$$

Hence prove that $R^2 = c^2 - (c - 4h)^2$.

If u is held fixed while α varies, so that R and h vary, deduce from the last equation that R is an increasing function of h when $h < c/4$.

Hence, prove that, if a particle is projected with speed 30 m s^{-1} from the floor of a horizontal tunnel of height 20 m, the greatest range which can be attained in the tunnel is about 89.4 m.   (L)

● Using x and y as displacements horizontally and vertically from the point of projection respectively,

$$x = (u \cos \alpha) t, \qquad y = (u \sin \alpha) t - \tfrac{1}{2} g t^2.$$

By conservation of energy,

$$\dot{y}^2 = u^2 \sin^2 \alpha - 2gy.$$

At the maximum height $\dot{y} = 0$ and $y = h$,

$$0 = u^2 \sin^2 \alpha - 2gh \qquad \text{or} \qquad 2h = \frac{u^2}{g} \sin^2 \alpha = c \sin^2 \alpha.$$

When $y = 0$, $x = R$ and $t = \dfrac{2u \sin \alpha}{g}$.

$$R = (u \cos \alpha) t = \frac{2u^2}{g} (\sin \alpha)(\cos \alpha) = 2c (\sin \alpha)(\cos \alpha).$$

Squaring,
$$
\begin{aligned}
R^2 &= 4c^2 \sin^2 \alpha \cos^2 \alpha \\
&= 4 [c \sin^2 \alpha][c(1 - \sin^2 \alpha)] \\
&= 4 (2h)(c - 2h) \\
&= 8ch - 16h^2 \\
&= c^2 - (c^2 - 8ch + 16h^2),
\end{aligned}
$$
$$R^2 = c^2 - (c - 4h)^2 . \tag{1}$$

Differentiating, $\qquad 2R \dfrac{\mathrm{d}R}{\mathrm{d}h} = 8(c - 4h), \qquad \dfrac{\mathrm{d}R}{\mathrm{d}h} > 0 \quad \text{when} \quad h < \dfrac{c}{4}.$

Thus R is an increasing function of h when $h < \dfrac{c}{4}$. $\tag{2}$

If $u = 30$, $\quad c = \dfrac{900}{10} = 90$. The maximum possible value of h is 20.

Since $h < \dfrac{c}{4}$, (2) is satisfied and the maximum possible range R is when $h = 20$.

Greatest range $= \sqrt{(90^2 - 10^2)} = 89.4$ m $\quad$ (from (1)).

7.4 A particle is projected under gravity with speed u and elevation $\dfrac{\pi}{4}$ from a point O on horizontal ground and hits the ground at P where $OP = R$. Taking the horizontal and vertical through O as axes of x and y respectively, show that the equation of the path is

$$y = x - \frac{gx^2}{u^2} \qquad \text{and that} \qquad R = \frac{u^2}{g} .$$

Q is the point on the path where $x = \dfrac{R}{4}$. Find

(a) the angle that OQ makes with the horizontal,
(b) the magnitude (in terms of u) and direction of the velocity of the particle at Q,
(c) the ratio of the times of flight from O to Q and Q to P. $\qquad$ (SUJB)

● Bookwork, from the Fact Sheet (section 7.1, equations (7) and (9)) with $\alpha = \dfrac{\pi}{4}$.

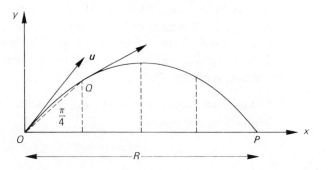

(a) At Q, $x = \dfrac{R}{4} = \dfrac{u^2}{4g}$.

Substituting into the equation of the path,

$$y = \frac{u^2}{4g} - \frac{(g)\,(u^4)}{(16g^2)\,(u^2)} = \frac{3u^2}{16g}.$$

The angle OQ makes with the horizontal is

$$\tan^{-1}\!\left(\frac{y}{x}\right) = \tan^{-1}\!\left(\frac{3}{4}\right) = 0.644 \text{ radians} \quad (\text{or } 36.9^\circ).$$

(b) By conservation of energy,

$$(\text{speed at } Q)^2 = u^2 - 2gy$$
$$= u^2 - \tfrac{3}{8}u^2 = \tfrac{5}{8}u^2.$$

From the equation of the path, $\dfrac{dy}{dx} = 1 - \dfrac{2gx}{u^2} = \tfrac{1}{2}$.

Thus velocity at Q is $(\sqrt{\tfrac{5}{8}})u$ at an angle of $\tan^{-1}\tfrac{1}{2}$ (0.464 radians or 26.6°) above the horizontal.

(c) Since the horizontal velocity is constant, the ratio of times of flight from O to Q and Q to P equals the ratio of the horizontal distances between the points, namely, $1 : 3$.

7.5 Two particles A and B are projected simultaneously under gravity; A from a point O on horizontal ground and B from a point 40 m vertically above O. B is projected horizontally with speed 28 m s^{-1}. If the particles hit the ground simultaneously at the same point, taking g as 9.8 m s^{-2}, calculate,

(a) the time taken for B to reach the ground and the horizontal distance it has then travelled;

(b) the magnitude and direction of the velocity with which A is projected.

Show that, just prior to hitting the ground, the directions of motion of A and B differ by about $18\tfrac{1}{2}^\circ$. (SUJB)

● Take the horizontal and vertically upward displacements from O as x and y respectively.

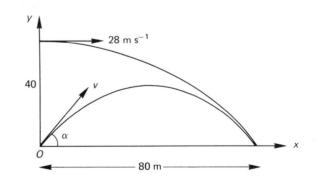

For B: $\dot{x}_B = 28$, $\dot{y}_B = -gt$.

$\qquad x_B = 28t$, $\quad y_B = 40 - \tfrac{1}{2}gt^2$.

When B strikes the ground $y_B = 0$.

so $t^2 = \dfrac{40}{4.9}$, $\quad t = \dfrac{20}{7}\,\text{s}$ (or 2.86 s).

At this time $x_B = 28t = 80$ m.

(a) B strikes the ground after 2.86 s having travelled 80 m horizontally.

For A: Let the horizontal and vertically upwards components of the velocity of projection be $v \cos \alpha$ and $v \sin \alpha$.

Then $\qquad x_A = (v \cos \alpha)\, t, \qquad y_A = (v \sin \alpha)\, t - \frac{1}{2}gt^2$.

Since A and B travel the same distance horizontally in the same time, in order to collide they must have the same horizontal speed i.e. $\quad v \cos \alpha = 28$. $\qquad (1)$

When $t = \dfrac{20}{7}$, $\quad y_A = 0$, so $\quad v (\sin \alpha)\, t = \frac{1}{2}gt^2$

$$v \sin \alpha = \frac{9.8}{2} \left(\frac{20}{7} \right) = 14. \qquad (2)$$

$$\text{and } \dot{y}_B = -9.8 \left(\frac{20}{7} \right) = -28.$$

Therefore, from (1) and (2), $\tan \alpha = \frac{1}{2}$, $\quad v = \sqrt{(14^2 + 28^2)} = 14\sqrt{5} \text{ m s}^{-2}$.

(b) The velocity of projection $= 31.3 \text{ m s}^{-1}$ at $26.6°$ to the ground.

Just before striking the ground, velocity of A makes the same angle as the angle of projection but below the horizontal,

and velocity of B makes an angle $\arctan \left(\dfrac{\dot{y}_B}{\dot{x}_B} \right) = 45°$ below the horizontal.

Therefore the directions of motion of A and B differ by about $18\frac{1}{2}°$.

7.6 A projectile is fired from a fixed point O at an elevation α, and hits a stationary target A at a distance a from O on the same level. Find the speed of projection.

On a second occasion the projectile is fired from O with the same speed and angle of projection as before. At the same instant the target is fired from its original position with speed v and elevation β in the plane of the path of the projectile in the direction away from O.

Prove that, if the projectile hits the target, it does so at a time

$$\frac{a}{v} \sin \alpha \operatorname{cosec} (\beta - \alpha)$$

after the projectile has been fired. $\hfill$ (OLE)

● Let the projectile have a speed of projection v_1 at an angle of elevation α.
Horizontal displacement $x = (v_1 \cos \alpha)\, t$. $\hfill (1)$
Vertical displacement $y = (v_1 \sin \alpha)\, t - \frac{1}{2}gt^2$. $\hfill (2)$
At the target $x = a$, $y = 0$.

From (2) $\quad t = \dfrac{2v_1 \sin \alpha}{g}$.

In (1) $\qquad a = \dfrac{2v_1^2 \sin \alpha \cos \alpha}{g}$ (this is the range).

Therefore speed of projection $v_1 = \sqrt{\left(\dfrac{ag}{\sin 2\alpha} \right)}$.

For the projectile to hit the target they must be in the same position at the same time.

To reach the same height in the same time the initial vertical components of velocity must be equal.

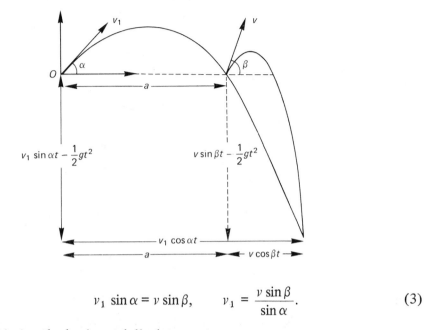

$$v_1 \sin \alpha = v \sin \beta, \qquad v_1 = \frac{v \sin \beta}{\sin \alpha}. \qquad (3)$$

Considering the horizontal displacements:

To reach the same horizontal position at the same instant, the projectile must travel a horizontal distance a more than that travelled by the target:

$$(v_1 \cos \alpha) t = (v \cos \beta) t + a.$$

Using (3), $\left(\dfrac{v \sin \beta \cos \alpha}{\sin \alpha} - (v \cos \beta) \right) t = a$

Multiplying by $\sin \alpha$,

$$v (\sin \beta \cos \alpha - \cos \beta \sin \alpha) t = a \sin \alpha.$$

Therefore $\qquad\qquad v \sin (\beta - \alpha) t = a \sin \alpha,$

or $\qquad\qquad\qquad\qquad t = \dfrac{a}{v} \sin \alpha \, \mathrm{cosec} \, (\beta - \alpha).$

7.7 A particle is projected from a point O at a time $t = 0$ with speed v and angle of elevation α. It moves under gravity and reaches its range R at time $t = T$. Show that

$$T = \frac{2v \sin \alpha}{g} \qquad \text{and} \qquad R = \frac{2v^2 \sin \alpha \cos \alpha}{g}.$$

When $t = \frac{1}{4}T$,

(a) find the height and the speed of the particle:

(b) If the velocity vector then makes an angle β with the horizontal, show that $\tan \beta = \frac{1}{2} \tan \alpha$.

Find the value of t in terms of T when the position vector makes an angle γ below the horizontal where $\tan \gamma = \frac{1}{3} \tan \alpha$.

● Bookwork: see Fact Sheet, section 7.1, equation (7).

At a general point $\quad \dot{x} = v \cos \alpha, \quad x = (v \cos \alpha) t;$
$$\dot{y} = v \sin \alpha - gt, \quad y = (v \sin \alpha) t - \tfrac{1}{2} gt^2.$$

When $t = \frac{1}{4}T = \dfrac{v \sin \alpha}{2g}$:

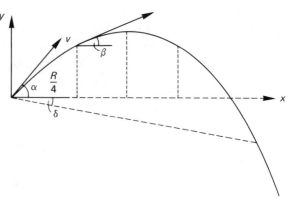

(a) Height $= \dfrac{v^2 \sin^2 \alpha}{2g} - \dfrac{gv^2 \sin^2 \alpha}{2 \cdot 4g^2} = \dfrac{3v^2}{8g} \sin^2 \alpha;$

$\dot{x} = v \cos \alpha, \quad \dot{y} = v \sin \alpha - \dfrac{v}{2} \sin \alpha = \dfrac{v}{2} \sin \alpha.$

Speed $= v\sqrt{(\cos^2 \alpha + \frac{1}{4} \sin^2 \alpha)} = \dfrac{v}{2}\sqrt{(4 - 3 \sin^2 \alpha)}.$

(b) $\tan \beta = \dfrac{\dot{y}}{\dot{x}} = \dfrac{v \sin \alpha}{2v \cos \alpha} = \frac{1}{2} \tan \alpha.$

Since γ is below the horizontal,

$$-\tan \gamma = \dfrac{y}{x} = \dfrac{(v \sin \alpha)\,t - \frac{1}{2}gt^2}{(v \cos \alpha)\,t} = \tan \alpha - \dfrac{gt}{2v \cos \alpha}.$$

If $\tan \gamma = \frac{1}{3} \tan \alpha,$ then $\frac{4}{3} \tan \alpha = \dfrac{gt}{2v \cos \alpha},$ $\quad t = \dfrac{8v}{3g} \sin \alpha = \frac{4}{3}T.$

Therefore the position vector makes an angle γ below the horizontal when $t = \frac{4}{3}T.$

7.8   An aircraft is flying with speed v in a direction inclined at an angle α above the horizontal. When the aircraft is at a height h, a bomb is dropped. Show that the horizontal distance R, measured from the point vertically below the point at which the bomb is dropped to the point where the bomb hits the ground, is given by

$$gR = \tfrac{1}{2}v^2 \sin 2\alpha + v\,(2gh + v^2 \sin^2 \alpha)^{1/2} \cos \alpha. \tag{L}$$

● Let x and y be the horizontal and vertically upwards displacements respectively. At the moment of release ($t = 0$) the bomb has velocity components

$$\dot{x} = v \cos \alpha, \qquad\qquad \dot{y} = v \sin \alpha.$$

At any time t $\quad \dot{x} = v \cos \alpha, \qquad\qquad \dot{y} = v \sin \alpha - gt,$

$$x = (v \cos \alpha)\,t, \qquad\qquad y = (v \sin \alpha)\,t - \tfrac{1}{2}gt^2.$$

When the bomb hits the ground $y = -h,$

$$\Rightarrow \quad -h = (v \sin \alpha)\,t - \tfrac{1}{2}gt^2 \quad (1), \qquad x = R = (v \cos \alpha)\,t \quad (2).$$

From (1), $\frac{1}{2}gt^2 - (v \sin \alpha)\,t - h = 0,$

$$t = \dfrac{v \sin \alpha \pm \sqrt{(v^2 \sin^2 \alpha + 2gh)}}{g}.$$

Since t is positive and $\sqrt{(v^2 \sin^2 \alpha + 2gh)} > v \sin \alpha$, the positive sign must be taken.

$$t = \dfrac{v \sin \alpha + \sqrt{(v^2 \sin^2 \alpha + 2gh)}}{g}.$$

In (2), $\quad R = \dfrac{v \cos \alpha \,[v \sin \alpha + \sqrt{(v^2 \sin^2 \alpha + 2gh)}]}{g}.$

Therefore $\quad gR = \tfrac{1}{2}v^2 \sin 2\alpha + v\,(2gh + v^2 \sin^2 \alpha)^{1/2} \cos \alpha.$

7.9 A small particle P is projected from a point O on a horizontal plane so that it first lands again at a point on the plane at a distance $2a$ from O. The maximum height reached by the particle is b and in this position the particle is directly above a point B on the plane. Find, in terms of a, b and g, the tangent of the angle of projection and the square of the speed of projection.

At time t the horizontal and vertical displacements of P from O are x and y respectively.

Show, using any relevant results from the book of formulae, that

$$y = \frac{2bx}{a} - \frac{bx^2}{a^2}$$

Hence determine, in terms of a and b, constants p, q and r such that $y = -p(x-q)^2 + r$, and so show that

$$BP^2 = \frac{b^2}{a^4}(x-a)^4 - 2\frac{b^2}{a^2}(x-a)^2 + (x-a)^2 + b^2.$$

Hence find the greatest value of b such that BP is never less than b throughout the motion. (AEB 1984)

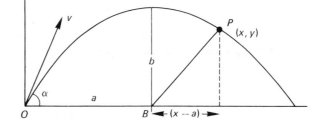

- Equation of trajectory $\quad y = x\tan\alpha - \dfrac{gx^2\sec^2\alpha}{2v^2}$.

At the highest point (a, b) the vertical velocity $\dot{y} = 0$.
Using $\quad \dot{y}^2 = (v\sin\alpha)^2 - 2gy$,

$$0 = v^2\sin^2\alpha - 2gb, \qquad \Rightarrow \qquad v^2\sin^2\alpha = 2gb. \tag{1}$$

Using $R = \dfrac{2v^2\sin\alpha\cos\alpha}{g}$,

$$2a = \frac{2v^2\sin\alpha\cos\alpha}{g} \qquad \Rightarrow \qquad v^2\sin\alpha\cos\alpha = ga. \tag{2}$$

(1)/(2) gives $\quad \tan\alpha = \dfrac{2b}{a} \qquad$ so $\sin^2\alpha = \dfrac{4b^2}{4b^2+a^2} \quad$ and $\quad \cos^2\alpha = \dfrac{a^2}{4b^2+a^2}$

In (1) $\quad v^2 = \dfrac{2gb}{\sin^2\alpha} = \dfrac{g(4b^2+a^2)}{2b} ; \qquad \left(2v^2\cos^2\alpha = \dfrac{ga^2}{b}\right)$.

In the trajectory equation $y = x\left(\dfrac{2b}{a}\right) - \dfrac{gx^2b}{ga^2}$

$$y = \frac{2bx}{a} - \frac{bx^2}{a^2}$$

$$= -\frac{b}{a^2}(x^2 - 2ax + a^2) + b$$

$$= -\frac{b}{a^2}(x-a)^2 + b.$$

Hence $\quad p = \dfrac{b}{a^2}, \quad q = a, \quad r = b.$

$$BP^2 = (x - a)^2 + y^2 = (x - a)^2 + \frac{b^2}{a^4}(x - a)^4 - \frac{2b^2}{a^2}(x - a)^2 + b^2$$

$$= \frac{b^2}{a^4}(x - a)^4 - 2\frac{b^2}{a^2}(x - a)^2 + (x - a)^2 + b^2$$

$$= b^2\left(\frac{x - a}{a}\right)^2\left[\left(\frac{x - a}{a}\right)^2 - 2 + \frac{a^2}{b^2}\right] + b^2.$$

If $BP \geqslant b$ for all x, $2 - \left(\frac{x - a}{a}\right)^2 \leqslant \frac{a^2}{b^2}$.

Minimum value of $(x - a)^2$ is zero so $\frac{a^2}{b^2} \geqslant 2$ or $\frac{b}{a} \leqslant \frac{1}{\sqrt{2}} \Rightarrow b \leqslant \frac{a}{\sqrt{2}}$.

7.10 The muzzle speed of a gun is v and it is desired to hit a small target at a horizontal distance a away and at a height b above the gun. Show that this is impossible if $v^2 (v^2 - 2gb) < g^2 a^2$, but that, if $v^2 (v^2 - 2gb) > g^2 a^2$, there are two possible elevations for the gun.

Show that if $v^2 = 2ag$ and $b = \frac{3}{4}a$, there is only one possible elevation, and find the time taken to hit the target. (OLE)

• Let the angle of elevation be α.

The trajectory equation is $y = x \tan \alpha - \frac{gx^2 (1 + \tan^2 \alpha)}{2v^2}$.

For the shell to hit the target at $x = a$, $y = b$,

$$b = a \tan \alpha - \frac{ga^2 (1 + \tan^2 \alpha)}{2v^2},$$

$$\Rightarrow \quad ga^2 \tan^2 \alpha - 2av^2 \tan \alpha + 2v^2 b + ga^2 = 0. \tag{1}$$

This is a quadratic equation in $\tan \alpha$, which must have real solutions for the target to be hit.

(a) The target cannot be hit if the discriminant < 0,

i.e. if
$$4a^2 v^4 - 4(ga^2)(2v^2 b + ga^2) < 0,$$
$$v^4 - 2gbv^2 - g^2 a^2 < 0$$
$$\Rightarrow \quad v^2 (v^2 - 2gb) < g^2 a^2.$$

(b) If the discriminant > 0, i.e. if $v^2 (v^2 - 2gb) > g^2 a^2$,
then equation (1) has two real distinct solutions for $\tan \alpha$ and hence for the elevation.

(c) If $v^2 = 2ga$ and $b = \frac{3}{4}a$,
then (1) becomes

$$ga^2 \tan^2 \alpha - 4ga^2 \tan \alpha + 3ga^2 + ga^2 = 0$$

$$\Rightarrow \quad \tan^2 \alpha - 4 \tan \alpha + 4 = 0 \quad \Rightarrow \quad \tan \alpha = 2,$$

i.e. only one possible elevation, arctan (2).

Horizontal displacement $a = (v \cos \alpha) t = \sqrt{(2ga)}\left(\frac{1}{\sqrt{5}}\right) t \Rightarrow t = \frac{a\sqrt{5}}{\sqrt{(2ga)}} = \sqrt{\left(\frac{5a}{2g}\right)}.$

Time taken for the shell to reach the target is $\sqrt{\left(\frac{5a}{2g}\right)}.$

7.3 Exercises

7.1 (multiple choice) A particle is projected with velocity 40 m s^{-1} at an angle of elevation arctan $\frac{3}{4}$. After 2 seconds the particle is moving at an angle θ to the horizontal.

Then tan θ =

A, $\frac{4}{3}$;   B, $\frac{16}{7}$;   C, $\frac{7}{16}$;   D, $\frac{3}{4}$;   E, $\frac{1}{8}$.

7.2 At time $t = 0$, a particle is projected from a point O with speed u at an angle of elevation α. At time t, the horizontal and vertical distances of the particle from O are x and y respectively. Express x and y in terms of u, α, t and g. Hence show that

$$y = x \tan \alpha - \frac{gx^2}{2u^2} (1 + \tan^2 \alpha).$$

A golf ball is struck from a point A, leaving A with speed 30 m s^{-1} at an angle of elevation θ, and lands, without bouncing, in a bunker at a point B, which is at the same horizontal level as A. Before landing in the bunker, the ball just clears the top of a tree which is at a horizontal distance of 72 m from A, the top of the tree being 9 m above the level of AB. Show that one of the possible values of θ is $\tan^{-1} \frac{3}{4}$ and find the other value. Given that θ was in fact $\tan^{-1} \frac{3}{4}$, find the distance AB. (Take g as 10 m s^{-2}). (L)

7.3 A tennis player hits a ball at a point O, which is at a height of 2 m above the ground and at a horizontal distance 4 m from the net, the initial speed being in a direction of 45° above the horizontal in a vertical plane perpendicular to the net. The ball just clears the net which is 1 m high.

(a) Taking the horizontal and vertical through O in the plane of motion as axes of x and y respectively, show that the equation of the path of the ball may be written in the form $y = x - \dfrac{5x^2}{16}$, assuming that the only force acting on the ball is that due to gravity.

Find the initial speed of the ball.

Find, also,

(b) the distance from the net at which the ball strikes the ground.
(c) the magnitude and direction of the velocity with which the ball strikes the ground.

All answers should be given correct to two significant figures. (SUJB)

7.4 O is a point on horizontal ground. D is a point vertically above O. A particle A is projected from O with a speed u m s^{-1} at an angle of elevation α. Simultaneously, a second particle B is projected horizontally from D with a speed v m s^{-1} on the same side of OD as A. Show that, if the particles collide, then $v = u \cos \alpha$. Find a second condition which must also be satisfied.

Given that $u = 51$, $v = 45$, $\tan \alpha = \frac{8}{15}$, and $d = 60$, satisfy these conditions, verify that a collision will occur.

Find (a) the position of the particles on collision, and (b) the speed of A just before the collision.

7.5 A vertical tower of height h stands on level ground. A gun on the top fires a shell with angle of elevation α. When the shell hits the ground its path makes an angle $\tan^{-1}(2\tan\alpha)$ with the horizontal. Prove that the speed of projection is $(\frac{2}{3}gh)^{1/2}\operatorname{cosec}\alpha$ and find the distance from the foot of the tower of the point where the shell lands. (OLE)

7.6 A particle is projected from a point O on a horizontal plane with velocity u, which makes an angle α with the plane. Unit vectors $\mathbf{i}$ and $\mathbf{j}$ are chosen such that the direction of $\mathbf{i}$ is horizontal and makes an angle α with the initial velocity, and $\mathbf{j}$ points vertically upwards. Given that the particle moves freely under gravity, find its position vector $\mathbf{r}$, relative to O, at time t after projection. Calculate $\mathbf{r}\cdot\mathbf{j}$ and $\mathbf{r}\cdot\mathbf{i}$ and hence deduce the time of flight and the range of the particle along the plane.

The particle is now fired with speed u at a target which is a horizontal distance D away from O. If the angle of projection, α, were $\pi/3$ the particle would fall short of the target by 10 m but if the angle of projection were $\pi/4$ the particle would overshoot the target by 10 m. Find D and a correct angle of projection for the particle to hit the target. (AEB 1983)

7.7 A particle is projected vertically upwards with speed v from a point O. Find H_O, the maximum height of the particle above O. If the particle, projected with the same velocity v but at an angle α with the horizontal, reaches a height of H, and has a range on the horizontal plane of R, find:
(a) H in terms of H_O and α,
(b) the value of α for which $H = 0.25H_O$,
(c) the value of R in terms of H_O for this value of α.
Find the value of α if $R = 2H$.

7.8 A particle is projected at time $t = 0$ from an origin O on a fixed horizontal plane with speed u at an angle α to the horizontal. The distances travelled horizontally and vertically by time t are x and y respectively.

Starting from the equations of motion

$$\frac{d^2x}{dt^2} = 0 \qquad \text{and} \qquad \frac{d^2y}{dt^2} = -g,$$

show, by integration, that $y = x\tan\alpha - \dfrac{gx^2\sec^2\alpha}{2u^2}$

and that the maximum value of y is $\dfrac{u^2\sin^2\alpha}{2g}$.

A stone is thrown from a point O on level ground with a speed 13 m s^{-1} at an angle $\tan^{-1}\frac{12}{5}$ to the horizontal.

The stone just misses the top of a pole in its path and then reaches a maximum height of twice the height of the pole. At the instant the stone is thrown, a bird on top of the pole sets off with constant speed v m s^{-1} away from O in a horizontal line in a vertical plane containing O and the pole.
Find
(a) the distance, in metres correct to one decimal place, of the base of the pole from O.
(b) v, correct to one decimal place, given that the stone hits the bird.

(AEB 1983)

7.9 Two parallel walls each of height $a/4$ are on level ground and are distant a and $2a$ from a point O on the ground. A ball is projected from O in a vertical plane perpendicular to the walls and just clears both walls. Find the magnitude and direction of the velocity with which it is projected. If P is the highest point of the ball's trajectory, find the angle that OP makes with the horizontal.   (SUJB)

7.4 Brief Solutions to Exercises

7.1 $\dot{x} = 32$, $\dot{y} = 4$, $\tan\theta = \dfrac{\dot{y}}{\dot{x}} = \frac{1}{8}$.   <u>Answer **E**</u>

7.2   Bookwork: see Fact Sheet, section 7.1, equation (a).
Substitute $x = 72, y = 9$ $\Rightarrow$ $\tan\theta = \frac{3}{4}$ or $\frac{7}{4}$.
$\tan\theta = \frac{3}{4}$, $\sin\theta = \frac{3}{5}$, $\cos\theta = \frac{4}{5}$,   gives range $R = 86.4$ m.

7.3 $x = \dfrac{vt}{\sqrt{2}}$ in $y = \dfrac{vt}{\sqrt{2}} - \frac{1}{2}gt^2$ gives $y = x - \dfrac{gx^2}{v^2}$.

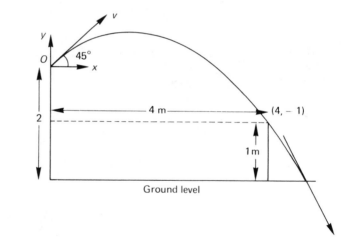

When $x = 4, y = -1$ so $v^2 = \dfrac{16g}{5} = 32$,

$\Rightarrow$ (a) initial speed $= \sqrt{\left(\dfrac{16g}{5}\right)} = 5.7$ m s^{-1} and $y = x - \dfrac{5x^2}{16}$.

   (b) When $y = -2$, $-2 = x - \dfrac{5x^2}{16}$ $\Rightarrow$ $x = 4.6$ m.

   Ball strikes the ground 0.6 m beyond the net.
   (c) Using Fact Sheet equation (6), $V^2 = 32 + 4g = 72$.
   Speed when ball strikes ground $= 8.5$ m s^{-1}.

   $\dfrac{dy}{dx} = 1 - \frac{5}{8}x$;   when $x = 4.6$, $\dfrac{dy}{dx} = -1.875 = \tan(-62°)$.

   Direction is $62°$ below the horizontal.

7.4 $\quad x_A = (u \cos \alpha) t, \quad y_A = (u \sin \alpha) t - \frac{1}{2} g t^2.$

$\qquad x_B = vt, \quad y_B = d - \frac{1}{2} g t^2.$

For collision, $\quad (u \cos \alpha) t = vt \quad$ and $\quad (u \sin \alpha) t - \frac{1}{2} g t^2 = d - \frac{1}{2} g t^2$

$\qquad\qquad\qquad \Rightarrow \quad v = u \cos \alpha \quad (1), \qquad$ and $\qquad (u \sin \alpha) t = d.$ $\qquad$ (2)

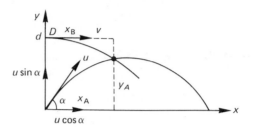

Given values satisfy (1), and (2) when $t = 2.5$.
(a) Collide at $x = 112.5$ m, $y = 28.75$ m.
(b) $\dot{x}_A = u \cos \alpha = 45$; when $t = 2.5$, $\dot{y}_A = u \sin \alpha - gt = -1$.
$\qquad$ Speed of A at that instant $= \sqrt{[(45)^2 + 1^2]} = 45.01$ m s^{-1}.

7.5 $\quad$ Use trajectory equation and its derivative.

When $y = -h$, $-h = x \tan \alpha - \dfrac{g x^2 \sec^2 \alpha}{2v^2}$ $\qquad\qquad\qquad\qquad\qquad$ (1)

and $-2 \tan \alpha = \tan \alpha - \dfrac{g x \sec^2 \alpha}{v^2} \quad \Rightarrow \quad \dfrac{g x \sec^2 \alpha}{v^2} = 3 \tan \alpha.$ $\qquad$ (2)

From (1) and (2), $\quad x = 2h \cot \alpha = $ required distance.
Substituting into (2), $\quad v = (\frac{2}{3} gh)^{1/2} \csc \alpha.$

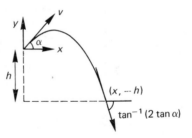

7.6 $\quad \ddot{\mathbf{r}} = -g\mathbf{j}, \quad \dot{\mathbf{r}} = u \cos \alpha \mathbf{i} + (u \sin \alpha - gt)\mathbf{j}, \quad \mathbf{r} = (u \cos \alpha) t \mathbf{i} + [(u \sin \alpha) t - \frac{1}{2} g t^2]\mathbf{j}.$
$\mathbf{r} \cdot \mathbf{j} = (u \sin \alpha) t - \frac{1}{2} g t^2 = $ height at time t;
$\mathbf{r} \cdot \mathbf{i} = (u \cos \alpha) t = $ horizontal displacement at time t.

$\mathbf{r} \cdot \mathbf{j} = 0$ gives time of flight, $\dfrac{2u \sin \alpha}{g}$.

Substitute into $\mathbf{r} \cdot \mathbf{i}$ to get range $R = \dfrac{2u^2 \sin \alpha \cos \alpha}{g}$.

When $\alpha = \dfrac{\pi}{3}$, $\quad D - 10 = \dfrac{\sqrt{3} u^2}{2g}$, $\quad$ when $\alpha = \dfrac{\pi}{4}$, $\quad D + 10 = \dfrac{u^2}{g}$.

Solving, $\quad D = 70 + 40\sqrt{3} = 139.3$ m, $\quad u^2 = 1493 \quad$ and $\quad \alpha = 34.4°$ or $55.6°$.

7.7 Projected vertically $\Rightarrow$ $v^2 = 2gH_O \Rightarrow H_O = \dfrac{v^2}{2g}$.

(a) Projected at α $\Rightarrow$ $v^2 \sin^2 \alpha = 2gH$ $\Rightarrow$ $H = H_O \sin^2 \alpha$.

(b) $H = \frac{1}{4}H_O$ $\Rightarrow$ $\sin \alpha = \frac{1}{2}$, $\alpha = 30°$.

(c) $R = \dfrac{2v^2 \sin \alpha \cos \alpha}{g} = \sqrt{3}H_O$.

$R = 2H$ $\Rightarrow$ $2\cos \alpha = \sin \alpha$ $\Rightarrow$ $\alpha = 63.4°$.

7.8 Bookwork: see Fact Sheet, section 7.1.

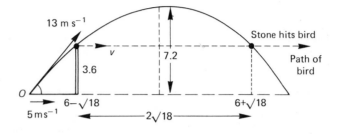

$\sin \alpha = \frac{12}{13}$, $\cos \alpha = \frac{5}{13}$. Maximum height = 7.2 m,
pole height = 3.6 m. When $y = 3.6$, $x = 3(2 \pm \sqrt{2})$.

(a) Pole distance = $3(2 - \sqrt{2})$.

(b) Hits bird when $x = 3(2 + \sqrt{2})$. Bird flies $6\sqrt{2}$ m.

$u \cos \alpha = 5$ m s^{-1}, time taken by stone = $\dfrac{3(2 + \sqrt{2})}{5}$ s.

Speed of bird = $10(\sqrt{2} - 1)$ m s^{-1}. = 4.14 m s^{-1}.

7.9 Substitute $\left(a, \dfrac{a}{4}\right)$, and $\left(2a, \dfrac{a}{4}\right)$ into trajectory
equation to get $\tan \alpha = \frac{3}{8}$ and $v = \frac{1}{4}\sqrt{(73ag)}$.

At highest point, $x = \frac{3}{2}a$, $y = \frac{9}{32}a$, so $\tan P\hat{O}X = \frac{3}{16}$. $P\hat{O}X = 10.6°$.

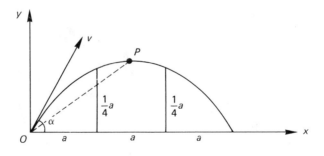

8 Motion in a Circle

Angular velocity. Motion in a circle with constant speed.
Proof that acceleration is $a\omega^2$. Motion in a vertical circle with constraining forces.

8.1 Fact Sheet

(a) Angular Velocity

The angular velocity ω of a particle P moving relative to a point O is defined $\omega = v/r$, where r is the distance OP and v the velocity of the particle perpendicular to OP.

(b) Circular Motion

(i) *Constant Speed* (*Vector derivation*)

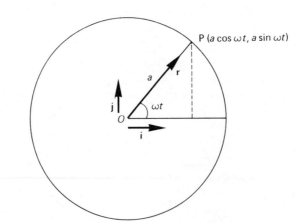

P moves in a circle, centre O, radius a with constant angular velocity ω.
Position $\overline{OP} = \mathbf{r} = a\cos\omega t\,\mathbf{i} + a\sin\omega t\,\mathbf{j}$, distance $OP = |\mathbf{r}| = a$.
Velocity $\dot{\mathbf{r}} = -a\omega\sin\omega t\,\mathbf{i} + a\omega\cos\omega t\,\mathbf{j}$,
speed $v = |\dot{\mathbf{r}}| = a\omega$, $\quad \mathbf{r} \cdot \dot{\mathbf{r}} = 0$ so $\mathbf{r}$ is perpendicular to $\dot{\mathbf{r}}$.
Acceleration $\ddot{\mathbf{r}} = -a\omega^2\cos\omega t\,\mathbf{i} - a\omega^2\sin\omega t\,\mathbf{j}$
$\qquad\qquad = -\omega^2\mathbf{r}$,
magnitude $= |\ddot{\mathbf{r}}| = a\omega^2 = v^2/a$, direction towards O.

Equations of motion

The sum of the components, towards the centre of the circle O, of the forces acting on $P = ma\omega^2 = mv^2/a$

$\Rightarrow \quad \Sigma F_{\text{towards } O} = ma\omega^2 = mv^2/a.$

Motion in a horizontal circle

(1) $\Sigma F_{\text{vertical}} = 0.$
(2) $\Sigma F_{\text{towards } O} = ma\omega^2 = mv^2/a.$

Satellites

The gravitational force acting on a particle, mass m, is km/d^2, where k is a constant and d is the distance from the particle to the centre of the planet.

When $d = a$ (= the radius of the planet), $mg = km/a^2$ where g is the acceleration due to the gravity experienced on the surface of the planet, giving $k = ga^2$.

Geostationary satellites above the earth have a period of 24 hours to enable them to remain above the same point on the equator.

(ii) **Variable Speed**

Motion in a vertical circle

In this case neither ω nor v will be constant. ω may be replaced by $\dot{\theta}$ and equation (2) becomes

(3) $\Sigma F_{\text{towards } O} = ma\dot{\theta}^2 = mv^2/a.$

(1) is replaced by the Equation of Conservation of Energy. (See Fact Sheet, section 7.1.)

8.2 Worked Examples

8.1 (multiple choice) A small ring of mass m is threaded on a thin smooth circular wire, which is of radius a and which is fixed with its plane vertical. The ring is projected from the lowest point of the wire with speed $\sqrt{(6ga)}$.

1 The speed of the ring at the highest point is $\sqrt{(2ga)}$.
2 The force exerted by the wire on the ring at the highest point is of magnitude $2mg$.
3 The force exerted by the wire on the ring at the highest point is directed upwards.

A, 1, 2 and 3 are correct;   B, only 1 and 2 are correct;
C, only 2 and 3 are correct;   D, only 1 is correct;
E, only 3 is correct.   (L)

- Kinetic energy (k.e.) at lowest point $= \frac{1}{2}m(6ga) = 3mga$.
   At highest point potential energy gained $= 2mga$.
   Thus k.e. at highest point $= 3mga - 2mga = mga = \frac{1}{2}mv^2$.
   Thus speed at highest point $= \sqrt{(2ga)}$:   (1) correct.

At highest point, accelerating force towards centre $= \dfrac{mv^2}{a} = \dfrac{m\,(2ga)}{a} = 2mg.$

Of this force, mg is due to gravity, so force exerted by wire $= 2mg - mg = mg$ downwards: (2) and (3) incorrect.   Answer **D**

8.2   A particle falling freely starts from rest at a point A which is 35 m above ground level. It eventually strikes the ground at B which is distant 40 m from a fixed point O on the ground. Obtain the angular speed of the particle about O when it has descended a distance 5 m. (Take g as 10 m s^{-2}.)   (L)

● When the particle has descended 5 m,

$$(\text{speed})^2 = 2g(5) = 100, \quad \Rightarrow \quad \text{speed} = 10 \text{ m s}^{-1}.$$

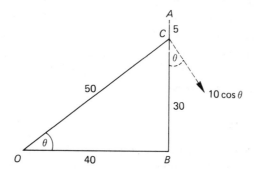

$\cos\theta = \frac{4}{5}$, so particle has a velocity $10\cos\theta = 8$ m s^{-1}, perpendicular to OC.
Distance of particle from O is 50 m.
Therefore angular speed $= \frac{8}{50}$ rad s^{-1} = 0.16 rad s^{-1}, in the sense of θ decreasing.

8.3   Two particles, A and B, of masses m_1 and m_2, are attached to the ends of a light inextensible string which passes over a smooth hook at O, which is free to rotate (see diagram). The particle A moves in a horizontal circle with constant speed v and B hangs at rest.
Find:
(a) OA,
(b) the angle AOB.

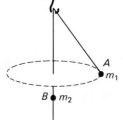

● Let angle $AOB = \alpha$, $OA = a$.
Then the radius of the circle $= a\sin\alpha$
For particle B:
Resolving vertically, $T = m_2 g.$ (1)

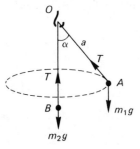

For particle A:
Resolving vertically, $T\cos\alpha = m_1 g.$ (2)

Eliminating T, $\cos\alpha = \dfrac{m_1}{m_2}.$

Therefore $\angle AOB = \alpha = \arccos\left(\dfrac{m_1}{m_2}\right).$

Resolving horizontally, $T\sin\alpha = \dfrac{m_1 v^2}{a\sin\alpha},$

i.e. $Ta\sin^2\alpha = m_1 v^2.$ (3)

But $\sin^2\alpha = 1 - \cos^2\alpha = 1 - \dfrac{m_1^2}{m_2^2}.$

Substituting in (3), $\quad m_2 g a \dfrac{(m_2^2 - m_1^2)}{m_2^2} = m_1 v^2 ,$

$$OA = a = \dfrac{m_1\, m_2\, v^2}{g(m_2^2 - m_1^2)}.$$

8.4 An artificial satellite of mass m moves under the action of a gravitational force which is directed towards the centre, O, of the earth and is of magnitude F. The orbit of the satellite is a circle of radius a and centre O. Obtain an expression for T, the period of the satellite, in terms of m, a and F.

Show that, if the gravitational force acting on a body of mass m at a distance r from O is $m\mu/r^2$, where μ is a constant, then $T^2 \mu = 4\pi^2 a^3$.

Assuming that the radius of the earth is 6400 km and that the acceleration due to gravity at a surface of the earth is 10 m s^{-2}, show that $\mu = (6.4)^2\ 10^{13}$ m^3 s^{-2}.

Hence, or otherwise, find the period of revolution, in hours to 2 decimal places, of the satellite when it travels in a circular orbit 600 km above the surface of the earth.

(L)

- Satellite moves in a circle radius a, angular velocity ω, so $F = m\omega^2 a$.

$$\omega = \sqrt{\left(\dfrac{F}{ma}\right)}.$$

Period of the satellite $T = \dfrac{2\pi}{\omega} = 2\pi \sqrt{\left(\dfrac{ma}{F}\right)}.$

If $F = \dfrac{m\mu}{r^2}$, and $a = r$, $T = 2\pi \sqrt{\left(\dfrac{ma^3}{m\mu}\right)}$, $\qquad$ i.e. $\mu T^2 = 4\pi^2 a^3$. $\hfill$ (1)

When $a = 6400 \times 10^3$, $F = mg = 10m$.

$$10m = \dfrac{m\mu}{6400^2 \times 10^6} \qquad \Rightarrow \qquad \mu = 10 \times 6400^2 \times 10^6 = (6.4)^2 \times 10^{13} \text{ m}^3 \text{ s}^{-2}.$$

From (1), $\qquad\qquad\qquad (6.4)^2 \times 10^{13}\, T^2 = 4\pi^2 a^3.$

When $a = (6400 + 600)$ km $= 7 \times 10^6$ m,

$$T^2 = \dfrac{4\pi^2 \times 7^3 \times 10^{18}}{(6.4)^2 \times 10^{13}} = 3.306 \times 10^7 ,$$

$$T = 5750 \text{ seconds}$$

$$= 1.60 \text{ hours.}$$

Period of revolution $= 1.60$ hours.

8.5 A particle of mass m is attached to one end A of a light string, and a particle of mass M is attached to the other end B. The string passes over a smooth hook, freely pivoted at a fixed point O, so that the end B with mass M hangs freely at rest while the end A with mass m moves in a horizontal circle with constant speed v.

Prove that

(a) OA is inclined at an angle $\cos^{-1} \dfrac{m}{M}$ to the vertical;

(b) the radius of the circle in which A moves is $\dfrac{\sqrt{(M^2 - m^2)}}{M}$ (OA);

(c) $v^2 = \dfrac{(M^2 - m^2)g}{Mm}$ (OA);

(d) the pressure on the hook is $\sqrt{2M(M + m)g}$. $\hfill$ (SUJB)

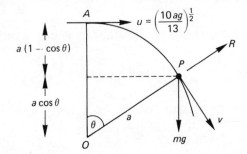

• Let OA be inclined at θ to the vertical.

For mass M, resolving vertically, $T = Mg$.

For mass m, resolving vertically, $T \cos \theta = mg$ $\Rightarrow$ $\cos \theta = \dfrac{m}{M}$,

thus (a) OA is inclined at an angle $\cos^{-1} \dfrac{m}{M}$ to the vertical.

Resolving horizontally, $\qquad T \sin \theta = \dfrac{mv^2}{r}$, $\qquad\qquad$ (1)

where $r = OA \sin \theta$.

Now $\sin \theta = \sqrt{(1 - \cos^2 \theta)} = \sqrt{\left(1 - \dfrac{m^2}{M^2}\right)} = \dfrac{\sqrt{(M^2 - m^2)}}{M}$.

(b) The radius of the circle $= \dfrac{\sqrt{(M^2 - m^2)}}{M} (OA)$.

From (1), $\dfrac{mv^2}{OA \sin \theta} = T \sin \theta$

$\Rightarrow \quad v^2 = \dfrac{T \sin^2 \theta}{m} (OA) = \dfrac{Mg(M^2 - m^2)}{mM^2} (OA)$

(c) $\quad v^2 = \dfrac{(M^2 - m^2)g}{Mm} (OA)$.

(d) The pressure on the hook $= 2T \cos \dfrac{\theta}{2} = 2Mg \cos \dfrac{\theta}{2}$.

$\cos \theta = 2 \cos^2 \dfrac{\theta}{2} - 1$, so $\cos^2 \dfrac{\theta}{2} = \dfrac{1 + \cos \theta}{2} = \dfrac{M + m}{2M}$.

Pressure on the hook $= 2Mg \sqrt{\left(\dfrac{M + m}{2M}\right)} = \sqrt{[2M(M + m)]}g$.

8.6 A smooth sphere, centre O, radius a, is fixed with its lowest point on a horizontal plane. A particle P of mass m is placed at the highest point A and given a tangential velocity of $(\frac{10}{13}ag)^{1/2}$. Show that the particle loses contact with the sphere when OP makes an angle $\arccos \frac{12}{13}$ with OA.

Find the angle which the direction of motion of the particle makes with the plane just before impact.

• When OP makes an angle θ with OA the particle has descended a distance $(a - a \cos \theta)$ and has a velocity v.

Hook

110

By conservation of energy,

$$\tfrac{1}{2}mv^2 = \tfrac{1}{2}m\,(\tfrac{10}{13}ag) + mga\,(1 - \cos\theta),$$

$$v^2 = \tfrac{10}{13}ag + 2ag\,(1 - \cos\theta) = \tfrac{36}{13}ag - 2ag\cos\theta.$$

Since the particle is moving in a circle the force towards the centre $= m\dfrac{v^2}{a}$.

So $\quad mg\cos\theta - R = mg\,(\tfrac{36}{13} - 2\cos\theta)$.

Particle loses contact with the sphere when $R = 0$.

Then $\quad \cos\theta = \tfrac{36}{13} - 2\cos\theta \quad \Rightarrow \quad \cos\theta = \tfrac{12}{13}$.

Particle loses contact with the sphere when OP makes an angle $\arccos\tfrac{12}{13}$ with OA.

At this point $v^2 = ag\,(\tfrac{36}{13} - \tfrac{24}{13}) = \tfrac{12}{13}ga$ and the particle is at a height $a(1 + \cos\theta)$ above the ground.

Vertical velocity component $= v\sin\theta$.

By conservation of energy (for projectiles), $\dot{y}^2 = v^2\sin^2\theta + 2gy$;

when the particle reaches the plane, $\dot{y}^2 = v^2\sin^2\theta + 2ga(1 + \cos\theta)$

$$\dot{y}^2 = \frac{12}{13}ag\left(\frac{25}{169}\right) + \frac{50}{13}ag = 3.983\,ag.$$

$\dot{x}^2$ (remains constant) $= v^2\cos^2\theta = 0.7865\,ag.$

Thus, just before impact, $\dfrac{\dot{y}}{\dot{x}} = 2.250.$

Just before impact direction of motion makes an angle α with the plane, where $\tan\alpha = 2.250$, $\alpha = 66.0°$.

8.7 A particle P of mass 200 g moving on a smooth horizontal plane with constant speed v m s^{-1} describes a circle centre O such that $OP = s$ m. The particle is subject to two forces, one towards O with magnitude $8v$ N and one away from O with magnitude λ/s^2 N.

(a) Given that $\lambda = 75$ and $s = 1$, find the possible values of v.

(b) If the period of revolution is $\dfrac{\pi}{5}$ when $v = 20$, find s and λ.

(c) Find the range of values of λ if $s = 1$.

• Force towards the centre of the circle $= 8v - \dfrac{\lambda}{s^2}$.

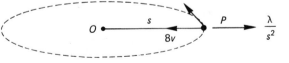

so $\qquad\qquad (0.2)\left(\dfrac{v^2}{s}\right) = 8v - \dfrac{\lambda}{s^2}.$ \hfill (1)

(a) When $s = 1$, $\lambda = 75$; $\qquad\qquad 0.2v^2 = 8v - 75$

$$v^2 - 40v + 375 = 0$$
$$(v - 25)\,(v - 15) = 0$$
$$v = 15 \text{ or } 25 \text{ m s}^{-1}.$$

(b) Period of revolution is $\dfrac{2\pi s}{v} = \dfrac{2\pi s}{20}$.

When the period is $\frac{\pi}{5}$, $s = 2$.

From (1), $\quad \dfrac{0.2\,(400)}{2} = 160 - \dfrac{\lambda}{4}$; $\quad \Rightarrow \quad \lambda = 480$.

(c) If $s = 1$, $\qquad 0.2v^2 = 8v - \lambda$

$$v^2 - 40v + 5\lambda = 0.$$

v must be real, so the discriminant $\geqslant 0$,

i.e. $\quad 1600 - 4\,(5\lambda) \geqslant 0, \quad 0 \leqslant \lambda \leqslant 80$.

8.8 A straight horizontal wire is free to rotate about a vertical axis through one end O. A bead of mass m is threaded on the wire and is attached by a light thread, passing through a small smooth ring at O, to a particle of mass $8m$ which is free to move in a vertical plane. The wire is made to rotate with constant angular speed ω about the vertical axis through O so that the hanging particle remains stationary and the bead remains at rest relative to the wire at distance a from O.

(a) Given that the wire is smooth calculate
  (i) the angular speed ω,
  (ii) the speed of the bead,
  (iii) the time for one complete rotation of the wire.
(b) Given that the wire is rough with coefficient of friction $\frac{1}{4}$, determine the permissible range of ω.
(c) Find the permissible range of ω if the wire is rough as in (b) and the bead is of mass $8m$, and the hanging particle of mass m. (AEB 1982)

● (a) Let the tension in the string be T.

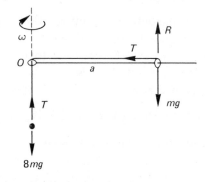

For the particle mass $8m$, resolving vertically, $\quad 8mg = T$.
For the bead mass m, $\quad T = m\omega^2 a$.

(i) $8mg = m\omega^2 a \quad \Rightarrow \quad \omega = 2\sqrt{\left(\dfrac{2g}{a}\right)}$.

(ii) Speed $= \omega a = 2\sqrt{(2ga)}$.

(iii) Time for one complete revolution $= \dfrac{2\pi}{\omega} = \pi\sqrt{\left(\dfrac{a}{2g}\right)}$.

(b) If the wire is rough.
As before $T = 8mg$.
For bead mass m:
Normal reaction on the bead $= mg$.

Maximum frictional force acting on the bead $= \dfrac{mg}{4}$.

Resolving horizontally. When the bead is about to move towards O,

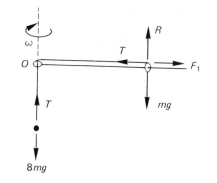

$$T - \frac{mg}{4} = m\omega^2 a \qquad \text{so} \qquad \frac{31}{4} mg = \omega^2 a \quad \Rightarrow \quad \omega = \sqrt{\left(\frac{31g}{4a}\right)}.$$

When the bead is about to move away from O, the frictional force is towards O.

$$T + \frac{mg}{4} = m\omega^2 a \quad \Rightarrow \quad \omega = \sqrt{\left(\frac{33g}{4a}\right)}.$$

Range of values of ω is $\sqrt{\left(\frac{31g}{4a}\right)} \leqslant \omega \leqslant \sqrt{\left(\frac{33g}{4a}\right)}$.

(c) If the bead has a mass $8m$ and the hanging particle a mass m then

$$T = mg \qquad \text{and} \qquad F_{\max} = 2mg \quad \text{(when about to slip)}.$$

When/if bead is about to move towards O, $\quad T - F = 8m\omega^2 a, \quad mg - F = 8m\omega^2 a.$
  Value of F required to prevent motion is mg, which is less than $F_{\max}$, so
  friction will always prevent movement towards O.
    When bead is about to move away from O, $\quad T + F = 8m\omega^2 a.$

With maximum F, $\quad 8m\omega^2 a = 3mg \quad \Rightarrow \quad \omega^2 = \frac{3g}{8a}$.

Range of values of ω is now $0 < \omega \leqslant \sqrt{\left(\frac{3g}{8a}\right)}$.

8.9 One end of a light inextensible string of length a is attached to a fixed point O which is at a height $\frac{a}{3}$ above a smooth horizontal table. A particle P of mass m is attached to the other end of the string and rests on the table with the string taut. The particle is projected so that it moves in a circle on the table with constant speed v.

Show that the tension in the string is $\frac{9mv^2}{8a}$.

Find in terms of m, g, a and v the reaction exerted on P by the table. Show that $v^2 \leqslant \frac{8ga}{3}$.

• Let the reaction exerted by the table be R.

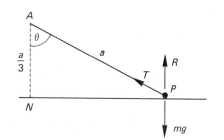

Resolving vertically, $T\cos\theta + R = mg.$ (1)

Resolving horizontally, $T\sin\theta = \dfrac{mv^2}{a\sin\theta}.$

But $\cos\theta = \dfrac{1}{3},\ \sin^2\theta = \dfrac{8}{9}$ so $T = \dfrac{9mv^2}{8a}.$

$$R = mg - \frac{9mv^2}{8a}\left(\frac{1}{3}\right) = m\left(g - \frac{3v^2}{8a}\right).$$

For the particle to remain in contact with the surface,

$$R \geqslant 0 \qquad \text{so} \qquad \frac{3v^2}{8a} \leqslant g \qquad \text{or} \qquad v^2 \leqslant \frac{8ag}{3}.$$

8.10 Show in a sketch the vector $\mathbf{i}\cos\theta - \mathbf{j}\sin\theta$, where $0 < \theta < \dfrac{\pi}{2}$ and $\mathbf{i}$ and $\mathbf{j}$ are unit vectors parallel to the x- and y-axes respectively. Deduce that this is a unit vector. Show on your diagram a unit vector perpendicular to this and express it in the form $a\mathbf{i} + b\mathbf{j}$.

A particle P describes a circle of radius $2r$ about the origin with constant angular speed 3ω, and, at time $t = 0$, P is at the point $(2r, 0)$ and is moving in the direction of decreasing y. Express $\overline{OP}$, at time t, in the form $a\mathbf{i} + b\mathbf{j}$.

State the magnitude of both the velocity and acceleration of P and the angles each of these vectors makes with $\overline{OP}$. Hence, or otherwise, express the velocity and acceleration of P in the form $a\mathbf{i} + b\mathbf{j}$.

A second particle Q has position vector given by $\overline{OQ} = r(\mathbf{i}\sin\omega t + \mathbf{j}\cos\omega t)$. Obtain, in its simplest possible form, an expression for PQ^2.   (AEB 1983)

- The magnitude of the vector $\mathbf{i}\cos\theta - \mathbf{j}\sin\theta = \sqrt{(\cos^2\theta + \sin^2\theta)}$
$$= 1.$$

A vector perpendicular to $\mathbf{i}\cos\theta - \mathbf{j}\sin\theta$ is $-\mathbf{i}\sin\theta - \mathbf{j}\cos\theta$.

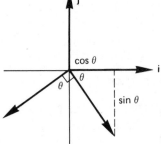

For P, $\mathbf{r} = \overline{OP} = 2r\cos3\omega t\,\mathbf{i} - 2r\sin3\omega t\,\mathbf{j}.$
Acceleration of $P = \ddot{\mathbf{r}} = -18\omega^2 r\cos3\omega t\,\mathbf{i} + 18\omega^2 r\sin3\omega t\,\mathbf{j}.$
Velocity $= \dot{\mathbf{r}} = -6\omega r\sin3\omega t\,\mathbf{i} - 6\omega r\cos3\omega t\,\mathbf{j}.$
$\dot{\mathbf{r}}$ has magnitude $6\omega r$ and makes an angle of $90°$ with $\overline{OP}$.
The magnitude of $\ddot{\mathbf{r}}$ is $18\omega^2 r$ and $\ddot{\mathbf{r}}$ makes an angle of $180°$ with $\overline{OP}$.

If $\overline{OQ} = r(\mathbf{i}\sin\omega t + \mathbf{j}\cos\omega t)$ then,
$\overline{PQ} = \mathbf{i}r(\sin\omega t - 2\cos3\omega t) + \mathbf{j}r(\cos\omega t + 2\sin3\omega t),$

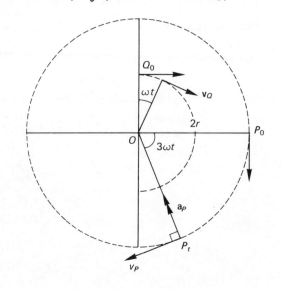

$$PQ^2 = r^2 (\sin \omega t - 2 \cos 3\omega t)^2 + r^2 (\cos \omega t + 2 \sin 3\omega t)^2$$
$$= r^2 (\sin^2 \omega t - 4 \sin \omega t \cos 3\omega t + 4 \cos^2 3\omega t)$$
$$+ r^2 (\cos^2 \omega t + 4 \cos \omega t \sin 3\omega t + 4 \sin^2 3\omega t)$$
$$= r^2 [5 + 4 (\cos \omega t \sin 3\omega t - \sin \omega t \cos 3\omega t)]$$
$$= r^2 (5 + 4 \sin 2\omega t).$$

8.3 Exercises

8.1 A particle moves with constant angular velocity ω in a horizontal circle of radius a on the inside of a fixed smooth hemispherical bowl of internal radius $2a$.

Show that $\omega^2 = \dfrac{g}{a\sqrt{3}}$.

8.2 A string has one end fixed at a point O with two particles of masses $2m$ and m attached at A and B respectively as shown in the diagram. The whole system is

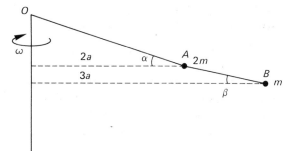

rotated with constant angular velocity ω about the vertical through O; the particles at A and B moving in horizontal circles of radii $2a$ and $3a$ respectively, and the sections OA and AB of the string making angles α and β with the horizontal.

If $\omega^2 = \dfrac{2g}{a}$, show that $\tan \alpha = \frac{3}{14}$ and find the tension in OA.

8.3 A particle is attached at one end of a light string, the other end of which is fixed. When the particle moves in a horizontal circle with speed 2 m s^{-1}, the string makes an angle $\tan^{-1} \frac{5}{12}$ with the vertical. Show that the length of the string is approximately 2.5 m.   (L)

8.4 A line OP of length r rotates in a plane about O with constant angular velocity ω. Prove that the acceleration of P is in the direction $\overline{PO}$ and has magnitude $r\omega^2$.

Two equal particles are attached to the ends A and B of a light inextensible string, which passes through a small hole at the apex C of a hollow, right, circular cone fixed with its axis vertical and apex uppermost. The semivertical angle of the cone is θ. The particle at A moves in a horizontal circle with constant angular velocity ω on the smooth surface of the cone, while the other particle hangs at rest inside the cone. If $CA = a$, prove that $\omega^2 = g/[a(1 + \cos\theta)]$ and deduce that

$$g/2\omega^2 < a < g/\omega^2.$$ (SUJB)

8.5 A particle A of mass m is attached to one end of a light inextensible string of length l. The other end is attached to a fixed point B on a rough vertical wire. A second string, also of length l, has one end attached to A and the other end to a small ring C of mass m which is threaded on the wire below B. The whole system rotates about the vertical BC with a constant angular velocity ω, with both strings taut. If $\angle ABC = \theta$ find the tensions in AB and AC, and show that $l\omega^2 > g \sec\theta$.

If $\theta = \sin^{-1}\frac{5}{13}$ and $l\omega^2 = 26g/9$ show that the coefficient of friction at C has a minimum possible value of $\frac{12}{25}$.

8.6 A particle of mass 0.1 kg moving on a smooth horizontal table with constant speed v m s^{-1} describes a circle centre O such that $OP = r$ m. The particle is attracted towards O by a force of magnitude $4v$ N and repelled from O by a force of magnitude (k/r) N where k is a constant.
(a) Given that $v = 40$ and the time of one revolution is $(\pi/10)$ s, find r and k.
(b) Given that $k = 30$ and $r = 1$, find the possible values of v.
(c) Find the range of values of k if $r = 1$. (AEB 1984)

8.7 An artificial satellite of mass m moves in a circular orbit around the earth under the action of a gravitational force F directed towards the centre of the earth. If $F = \dfrac{m\mu}{r^2}$, where r is the distance of the satellite from the centre of the earth, obtain an expression for the period of revolution of the satellite. Hence find the period of revolution, in hours, to one decimal place, of a satellite travelling in a circular orbit $36\,200$ km above the surface of the earth. (Assume that the radius of the earth is 6400 km and that $g = 10$ m s^{-2}.)

8.8 One end of a light inextensible string AB of length $2a + b$, is attached to a point A distance $\dfrac{a}{2}$ vertically above a point C on a horizontal table. A particle of mass m is attached to the point P of the string where $AP = a$. End B of the string is threaded down through a small smooth hole in the table at C and a mass $\dfrac{m}{2}$ is then attached at B.

The mass at B is too large to pass through the hole in the table which has a thickness b.

The particle at P is made to rotate about AC in a horizontal circle with constant angular velocity ω so that the particle at B is in contact with the underside of the table.

Find the tensions in both parts of the string and show that $a\omega^2 \geqslant 5g$.

Find the least tension the string has to sustain for this motion to be possible.

8.9 The position vector of a particle P of mass 250 g relative to the origin O is $(3\cos 2t\mathbf{i} + 3\sin 2t\mathbf{j})$ m at time t s.

Show that P is travelling in a circle with constant speed. Find the force acting on P at time t and show that this force is perpendicular to the velocity for all time t.

Find the maximum and minimum distances of P from the point A with position vector $3\sqrt{2}\mathbf{i} + 3\sqrt{2}\mathbf{j}$.

8.4 Brief Solutions to Exercises

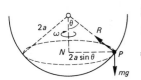

8.1 $\sin\theta = \frac{1}{2} \quad \Rightarrow \quad \theta = 30°$.

r.v.,* $R\cos\theta = mg$; r.h.,* $R\sin\theta = ma\omega^2$,

$\Rightarrow \quad \dfrac{2mg}{\sqrt{3}} = 2ma\omega^2, \quad \Rightarrow \quad \omega^2 = \dfrac{g}{a\sqrt{3}}$.

8.2 For B, r.v., $T_1\sin\beta = mg$, r.h., $T_1\cos\beta = m\omega^2(3a)$,

$\Rightarrow \quad \tan\beta = \dfrac{g}{3a\omega^2} = \dfrac{1}{6}$.

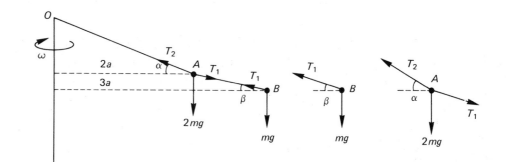

For A, r.v., $T_2\sin\alpha - T_1\sin\beta = 2mg$,

but $\qquad\qquad T_1\sin\beta = mg \quad \Rightarrow \quad T_2\sin\alpha = 3mg$. (1)

r.h., $T_2\cos\alpha - T_1\cos\beta = 2m\omega^2(2a)$,

but $\qquad\qquad T_1\cos\beta = 3ma\omega^2 \quad \Rightarrow \quad T_2\cos\alpha = 7ma\omega^2$. (2)

(1)/(2) gives $\qquad\qquad \tan\alpha = \dfrac{3mg}{7ma\omega^2} = \dfrac{3}{14} \quad \Rightarrow \quad T_2 = mg\sqrt{205}$.

8.3 $\tan\theta = \frac{5}{12} \quad \Rightarrow \quad \sin\theta = \frac{5}{13}$;

r.v., $T\cos\theta = mg$ (1); r.h., $T\sin\theta = \dfrac{mv^2}{r} = \dfrac{4m}{l\sin\theta}$ (2).

(1)/(2) gives $\quad \cot\theta = \dfrac{gl\sin\theta}{4} \quad \Rightarrow \quad \frac{12}{5} = \frac{10}{4}(l)\frac{5}{13} \quad$ so $\quad l \approx 2.5$ m.

8.4 Bookwork: see Fact Sheet, section 8.1(b).

For B, r.v., $\qquad\qquad T = mg$. (1)

For A, r.v., $\qquad\qquad T\cos\theta + R\sin\theta = mg$

$\qquad\qquad\qquad\qquad\qquad R\sin\theta = mg(1 - \cos\theta)$. (2)

r.h., $\qquad\qquad T\sin\theta - R\cos\theta = m\omega^2 a\sin\theta$

$\qquad\qquad\qquad\qquad\qquad R\cos\theta = m\sin\theta(g - \omega^2 a)$. (3)

*r.v. ≡ resolve vertically, r.h. ≡ resolve horizontally.

117

Eliminate R: $\qquad\qquad \sin^2\theta\,(g - \omega^2 a) = g\cos\theta\,(1 - \cos\theta)$,

$\qquad\qquad$ But $\qquad\qquad\qquad \sin^2\theta \equiv (1 - \cos\theta)\,(1 + \cos\theta)$.

Cancelling $(1 - \cos\theta)$,

$$(1 + \cos\theta)(g - \omega^2 a) = g\cos\theta \quad\Rightarrow\quad \omega^2 = \frac{g}{a\,(1 + \cos\theta)}.$$

$a = \dfrac{g}{\omega^2\,(1 + \cos\theta)}$ has max. value $\dfrac{g}{\omega^2}$ at $\theta = 90°$

$\qquad\qquad\qquad\qquad\qquad$ and min. value $\dfrac{g}{2\omega^2}$ at $\theta = 0°$.

Since $\theta \neq 0°$ or $90°$, $\quad \dfrac{g}{2\omega^2} < a < \dfrac{g}{\omega^2}$.

8.5 For A, r.v., $\qquad (T_1 - T_2)\cos\theta = mg \qquad$ or $\qquad T_1 - T_2 = mg\sec\theta$.

$\qquad\qquad$ r.h., $\qquad (T_1 + T_2)\sin\theta = m\omega^2 l\sin\theta \qquad T_1 + T_2 = m\omega^2 l$.

(1) + (2): $\qquad\qquad 2T_1 = m\,(g\sec\theta + \omega^2 l)$.

$$T_1 = \frac{m}{2}\,(\omega^2 l + g\sec\theta), \qquad T_2 = \frac{m}{2}\,(\omega^2 l - g\sec\theta);$$

$\qquad T_2 > 0 \quad$ so $\quad lw^2 > g\sec\theta$.

For C, r.h., $\qquad\qquad R = T_2\sin\theta = \dfrac{m}{2}\,(\omega^2 l\sin\theta - g\tan\theta)$.

$\qquad\qquad$ r.v., $\qquad\qquad F = mg - T_2\cos\theta = \dfrac{m}{2}\,(3g - \omega^2 l\cos\theta)$.

Substitute for $l\omega^2$ and θ.

$R = \dfrac{25}{72}\,mg, \quad F = \dfrac{mg}{6}; \quad \mu \geqslant \dfrac{F}{R} = \dfrac{12}{25} \quad\Rightarrow\quad \mu \geqslant \dfrac{12}{25}$.

8.6 $\qquad\qquad\qquad\qquad\qquad \dfrac{mv^2}{r} = 4v - \dfrac{k}{r}.$ $\qquad\qquad\qquad$ (1)

(a) Period $= \dfrac{2\pi r}{v}, \quad \dfrac{2\pi r}{40} = \dfrac{\pi}{10}, \quad\Rightarrow\quad r = 2$.

$\qquad$ In (1), $\quad \dfrac{0.1\,(40)^2}{2} = 160 - \dfrac{k}{2} \quad\Rightarrow\quad k = 160$.

(b) In (1), $\quad 0.1v^2 = 4v - 30, \quad v^2 - 40v + 300 = 0$,
$\qquad v = 10$ or 30.

(c) In (1), $\quad 0.1v^2 = 4v - k, \quad v^2 - 40v + 10k = 0$.
$\qquad v$ must be real. $\quad (40)^2 - 40k \geqslant 0, \quad k \leqslant 40$.
$\qquad k$ must be positive. $\quad 0 < k \leqslant 40$.

8.7 $m\omega^2 r = \dfrac{m\mu}{r^2}$.   Period $T = \dfrac{2\pi}{\omega} = 2\pi \sqrt{\left(\dfrac{r^3}{\mu}\right)}$.

On surface, $\dfrac{m\mu}{a^2} = mg \quad \Rightarrow \quad \mu = ga^2 \quad \Rightarrow \quad T = 2\pi \sqrt{\left(\dfrac{r^3}{ga^2}\right)}$.

Put $r = 42\,600 \times 10^3$, $a = 6400 \times 10^3$, $g = 10$.
Remember to change time from seconds to hours!

$$T = 23.98 \text{ h} = 24 \text{ h (a geostationary satellite)}.$$

8.8

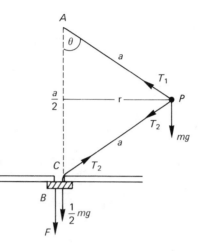

At P,   r.v., $(T_1 - T_2)\cos\theta = mg \quad \Rightarrow \quad T_1 - T_2 = 4mg.$ (1)

r.h., $(T_1 + T_2)\sin\theta = m\omega^2 a \sin\theta$,

$\Rightarrow \quad T_1 + T_2 = m\omega^2 a.$ (2)

(1) + (2): $2T_1 = m\,(a\omega^2 + 4g)$,

$T_1 = \dfrac{m}{2}\,(a\omega^2 + 4g)$,   (3) $T_2 = \dfrac{m}{2}\,(a\omega^2 - 4g)$.   (4)

At B,   force exerted by the table on particle $= F$.

r.v., $F + \dfrac{mg}{2} = T_2$, $F = \dfrac{m}{2}\,(a\omega^2 - 5g)$.

$F \geqslant 0$ since particle is in contact with table, $\Rightarrow \quad a\omega^2 \geqslant 5g$.

$T_1 \geqslant \dfrac{m}{2}\,(5g + 4g) = \dfrac{9}{2}\,mg$, $T_2 \geqslant \dfrac{m}{2}\,(5g - 4g) = \dfrac{mg}{2}$ $\Rightarrow$ least tension $= \dfrac{9}{2}\,mg$.

8.9 $\mathbf{r} = 3 \cos 2t\mathbf{i} + 3 \sin 2t\mathbf{j}$ $\Rightarrow$ $|\mathbf{r}| = 3\sqrt{(\cos^2 2t + \sin^2 2t)} = 3$.

Since $|\mathbf{r}|$ is constant, path is a circle, radius 3.

$\dot{\mathbf{r}} = -6 \sin 2t\mathbf{i} + 6 \cos 2t\mathbf{j}$, $|\dot{\mathbf{r}}| = 6$, so speed is constant.

$\ddot{\mathbf{r}} = -12 \cos 2t\mathbf{i} - 12 \sin 2t\mathbf{j} = -4\mathbf{r}$.

$\mathbf{F} = m\ddot{\mathbf{r}} = -3 (\cos 2t\mathbf{i} + \sin 2t\mathbf{j}) = -\mathbf{r}$,

$\Rightarrow$ force is 3 N towards centre of circle.

$\mathbf{F} \cdot \dot{\mathbf{r}} = 0$, so the force is perpendicular to velocity for all t.

Positions of P which are maximum and minimum distances from A are $\pm (\frac{3}{2}\sqrt{2}\mathbf{i} + \frac{3}{2}\sqrt{2}\mathbf{j})$, PA (min.) = 3, PA (max.) = 9.

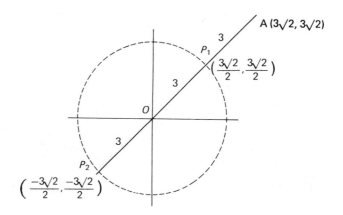

9 Differential Equations

Non-uniform acceleration involving differential equations of the types

$$\frac{dv}{dt} = f_1(t), \qquad \frac{dv}{dt} = f_2(v), \qquad v\frac{dv}{dx} = f_3(v) \qquad \text{and} \qquad v\frac{dv}{dx} = f_4(x).$$

Questions may involve the setting up of the equation of motion.

9.1 Fact Sheet

(a) Equations of Motion

For straight-line motion $F = m\ddot{x}$ so $\ddot{x} = \dfrac{F}{m}$,

where $\ddot{x} = \dfrac{d^2x}{dt^2} = \dfrac{dv}{dt} = \dfrac{dv}{dx}\dfrac{dx}{dt} = v\dfrac{dv}{dx} = \dfrac{d}{dx}(\tfrac{1}{2}v^2)$

$\left(x = \text{distance}, \ t = \text{time and } v = \dfrac{dx}{dt} = \text{speed}\right)$.

(b) Differential Equations

Decide which type of differential equation you have or which form of answer you require.

(i) $\dfrac{dv}{dt} = f_1(t) \quad \Rightarrow \quad \displaystyle\int dv = \int f_1(t)\,dt \quad \Rightarrow \quad v = F_1(t),$

(ii) $\dfrac{dv}{dt} = f_2(v) \quad \Rightarrow \quad \displaystyle\int \frac{dv}{f_2(v)} = \int dt \quad \Rightarrow \quad t = F_2(v),$

(iii) $v\dfrac{dv}{dx} = f_3(v) \quad \Rightarrow \quad \displaystyle\int \frac{v\,dv}{f_3(v)} = \int dx \quad \Rightarrow \quad x = F_3(v),$

(iv) $v\dfrac{dv}{dx} = f_4(x) \quad \Rightarrow \quad \displaystyle\int v\,dv = \int f_4(x)\,dx \quad \Rightarrow \quad \frac{v^2}{2} = F_4(x).$

In each case, putting $v = \dfrac{dx}{dt}$ in the resulting equation and integrating again will give a relationship between x and t.

For integration, see *Work Out Pure Mathematics 'A' Level* Chapters 16 and 17.

(c) Arbitrary Constants

Each solution of a differential equation includes an unknown, or arbitrary constant. It may be found either by substitution of the given values or by expressing the integrals as definite integrals.

Example $\dfrac{dv}{dt} = 5\omega \sin \omega t$ given that $v = 5$ when $t = 0$.

Either $\displaystyle\int dv = \int 5\omega \sin \omega t \, dt \quad \Rightarrow \quad v = -5 \cos \omega t + C;$

$v = 5$ when $t = 0$ so $5 = -5 + C, \quad \Rightarrow \quad C = 10, \quad v = 10 - 5 \cos \omega t;$

or $\displaystyle\int_{5}^{v} dv = \int_{0}^{t} 5\omega \sin \omega t \, dt \quad \Rightarrow \quad \Big[v \Big]_{5}^{v} = \Big[-5 \cos \omega t \Big]_{0}^{t}$

$\Rightarrow \quad v - 5 = -5 \cos \omega t + 5 \quad \Rightarrow \quad v = 10 - 5 \cos \omega t.$

(d) Second-order Differential Equations

$$a\dfrac{d^2 y}{dx^2} + b\dfrac{dy}{dx} + cy = ex + f.$$

Complementary function (C.F.)

The auxiliary equation is $am^2 + bm + c = 0$, roots m_1 and m_2.
 (i) Roots real and different, C.F. $= Ae^{m_1 x} + Be^{m_2 x}$.
 (ii) Roots real and equal,   C.F. $= e^{mx} (A + Bx)$.
(iii) Roots complex, $p \pm iq$,   C.F. $= e^{px} (A \cos qx + B \sin qx)$.

Particular integral (P.I.)

Assume a P.I. $= Ex + F$. Differentiate and substitute into the different equation to find E and F.

General solution: $y = $ C.F. $+$ P.I.

9.2 Worked Examples

9.1   A particle starts with speed 20 m s^{-1} and moves in a straight line. The particle is subjected to a resistance which produces a retardation which is initially 5 m s^{-2} and which increases uniformly with the distance moved, having a value of 11 m s^{-2} when the particle has moved a distance of 12 m. Given that the particle has speed $v \text{ m s}^{-1}$ when it has moved a distance of x m, show that, while the particle is in motion,

$$v \frac{dv}{dx} = -\left(5 + \frac{x}{2} \right).$$

Hence, or otherwise, calculate the distance moved by the particle in coming to rest.

(L)

- Retardation $= 5 + \lambda x$.

  When $x = 12$, retardation $= 11$ m s^{-2} so $\lambda = \frac{1}{2}$.

  Acceleration $= v\,\dfrac{\mathrm{d}v}{\mathrm{d}x} = -\left(5 + \dfrac{x}{2}\right)$.

  $$\int_{20}^{0} -v\,\mathrm{d}v = \int_{0}^{x_1}\left(5 + \frac{x}{2}\right)\mathrm{d}x,$$

  where x_1 is the distance moved in metres in coming to rest.

  $$-\left[\frac{v^2}{2}\right]_{20}^{0} = \left[5x + \frac{x^2}{4}\right]_{0}^{x_1}.$$

  $$\frac{400}{2} = 5x_1 + \frac{x_1^2}{4}.$$

  $$x_1^2 + 20x_1 - 800 = 0$$

  $$(x_1 - 20)(x_1 + 40) = 0.$$

  Required answer must be positive so the particle moves 20 m in coming to rest.

9.2 The motion of a particle satisfies the differential equation

$$\frac{\mathrm{d}^2 x}{\mathrm{d}t^2} - 9x = 27,$$

where x is its displacement from the origin at time t. Find a general solution for x.

Given that $x = 3$ and $\dfrac{\mathrm{d}x}{\mathrm{d}t} = -18$, when $t = 0$, find the particular solution and deduce the final displacement of the particle from O.

Find the time at which $x = 0$.

- $\dfrac{\mathrm{d}^2 x}{\mathrm{d}t^2} - 9x = 0$ has an auxiliary equation $m^2 - 9 = 0$, $m = \pm 3$, so complementary function is $A\mathrm{e}^{3t} + B\mathrm{e}^{-3t}$.

  Particular integral

  Try $x = C$. Substituting into the differential equation gives $0 - 9C = 27$, $C = -3$.

  General solution is $x = A\mathrm{e}^{3t} + B\mathrm{e}^{-3t} - 3$.

  Differentiating, $\dot{x} = 3A\mathrm{e}^{3t} - 3B\mathrm{e}^{-3t}$.

  When $t = 0$, $x = 3$, so $A + B - 3 = 3$ or $\qquad A + B = 6$;

  when $t = 0$, $\dfrac{\mathrm{d}x}{\mathrm{d}t} = -18$, so $3A - 3B = -18$ or $\qquad A - B = -6$.

  Adding, $A = 0$ so $B = 6$ and the particular solution is $x = 6\mathrm{e}^{-3t} - 3$.

  As $t \to \infty$, $x \to -3$; the final displacement is -3 units from O.

  When $x = 0$, $6\mathrm{e}^{-3t} = 3$, $\mathrm{e}^{-3t} = \frac{1}{2}$, $\mathrm{e}^{3t} = 2$, giving $t = \frac{1}{3}\ln 2 = 0.231$.

  Therefore $x = 0$ after a time 0.231.

9.3 A particle D moving along the x-axis has an acceleration in the positive x-direction of $k(26a^2 x - 8x^3)$ where k and a are positive constants. Given that D has a speed of $a^2\sqrt{(6k)}$ when $x = a\sqrt{3}$ obtain an expression for the speed of D for any value of x. Determine the values of x at which D comes instantaneously to rest and show that the motion of D is confined to a finite region of the x-axis.

(AEB 1983)

● $v \dfrac{dv}{dx} = k(26a^2 x - 8x^3).$

Separating and integrating,

$$\int_{a^2\sqrt{(6k)}}^{v} v\, dv = k \int_{a\sqrt{3}}^{x} (26a^2 x - 8x^3)\, dx,$$

$$\left[\frac{v^2}{2}\right]_{a^2\sqrt{(6k)}}^{v} = \left[k(13a^2 x^2 - 2x^4)\right]_{a\sqrt{3}}^{x}.$$

$$\frac{v^2}{2} - \frac{6}{2}ka^4 = 13ka^2 x^2 - 2kx^4 - 39ka^4 + 18ka^4,$$

$$\frac{v^2}{2} = 13ka^2 x^2 - 2kx^4 - 18ka^4,$$

$$v^2 = -2k(2x^4 - 13a^2 x^2 + 18a^4),$$

$$v = \sqrt{[-2k(2x^2 - 9a^2)(x^2 - 2a^2)]}.$$

$$v = 0 \quad \text{when} \quad x^2 = \frac{9}{2}a^2 \;\Rightarrow\; x = \pm\frac{3a}{\sqrt{2}},$$

$$\text{and when} \quad x^2 = 2a^2 \;\Rightarrow\; x = \pm\sqrt{2}a.$$

Initially, $x = a\sqrt{3}$, also $v^2 \geqslant 0$ requires $(2x^2 - 9a^2)(x^2 - 2a^2) \leqslant 0$,
hence $\sqrt{2}a \leqslant x \leqslant \dfrac{3a}{\sqrt{2}}$.

9.4 A car of mass 1200 kg moves along a horizontal road against a constant resistance of 2000 N. The engine is working at a constant rate of 70 kW. Find the time, to the nearest 0.1 of a second, for the speed to increase from 5 m s^{-1} to 25 m s^{-1}.

●
$$\text{Power} = \text{tractive force} \times \text{speed},$$

$$\text{tractive force} = \frac{70\,000}{v}\ \text{N},$$

$$\text{accelerating force} = \left(\frac{70\,000}{v}\right) - 2000\ \text{N},$$

$$\text{acceleration} = \frac{1}{1200}\left(\frac{70\,000}{v} - 2000\right) = \frac{175 - 5v}{3v}$$

$$\frac{dv}{dt} = \frac{5(35 - v)}{3v}.$$

Separate the variables and integrate:

$$\int_{5}^{25} \frac{3v}{35 - v}\, dv = \int_{0}^{t_1} 5\, dt.$$

Now, $\dfrac{v}{35 - v} = -1 + \dfrac{35}{35 - v}.$

Thus
$$3\left[-v - 35\ln(35 - v)\right]_{5}^{25} = \left[5t\right]_{0}^{t_1},$$

$$3[-25 - 35\ln 10 + 5 + 35\ln 30] = 5t_1,$$

$$t_1 = \tfrac{3}{5}(35\ln 3 - 20) = 11.1.$$

The speed increases from 5 m s^{-1} to 25 m s^{-1} in 11.1 seconds.

9.5 A particle moves in a straight line Ox and its velocity at time t is v when its displacement from O is x. Show that the acceleration of the particle can be expressed in the form $v\,\dfrac{dv}{dx}$.

The particle moves under the influence of a force proportional to $\dfrac{1}{x^3}$ and directed away from O.

Given that $x = a$, $v = 0$ when $t = 0$,

(a) show, by making a suitable choice of constant, that $3x^2v^2 = 4u^2(x^2 - a^2)$.

(b) Find an expression for x in terms of t.

● Acceleration $= \dfrac{dv}{dt} = \left(\dfrac{dv}{dx}\right)\left(\dfrac{dx}{dt}\right) = v\,\dfrac{dv}{dx}$.

(a) Force $= \dfrac{km}{x^3}$ where km is the constant of proportion.

$$mv\,\frac{dv}{dx} = \frac{mk}{x^3}, \quad \text{giving} \quad \int_0^v v\,dv = k\int_a^x \frac{1}{x^3}\,dx,$$

$$\frac{v^2}{2} = \frac{-k}{2x^2} + \frac{k}{2a^2},$$

$$\Rightarrow \quad v^2 x^2 = \frac{k}{a^2}(x^2 - a^2).$$

Putting $\dfrac{k}{a^2} = \dfrac{4}{3}u^2$, this becomes $3v^2x^2 = 4u^2(x^2 - a^2)$.

(b) If $\quad v^2 = \dfrac{4u^2}{3x^2}(x^2 - a^2) \quad$ then $v = \dfrac{dx}{dt} = \dfrac{2u\sqrt{(x^2 - a^2)}}{\sqrt{3}\,x}$.

Integrating, $$\int_a^x \frac{x\,dx}{\sqrt{(x^2 - a^2)}} = \int_0^t \frac{2u}{\sqrt{3}}\,dt,$$

$$\left[\sqrt{(x^2 - a^2)}\right]_a^x = \left[\frac{2ut}{\sqrt{3}}\right]_0^t,$$

i.e. $$\sqrt{(x^2 - a^2)} = \frac{2ut}{\sqrt{3}},$$

$$x^2 = \frac{4}{3}u^2 t^2 + a^2,$$

$$x = \sqrt{\left(\frac{4}{3}u^2 t^2 + a^2\right)}.$$

9.6 A particle is projected vertically upwards with an initial speed u through a medium which has a resistance mkv where m is the mass of the particle, k a constant and v the speed.

Find, in terms of u and k;

(a) the maximum height attained by the particle,

(b) the speed of the particle at time t,

(c) the time the particle takes to reach the maximum height.

- (a) $v \dfrac{\mathrm{d}v}{\mathrm{d}x} = -(g + kv)$.

Separate and integrate: $\displaystyle\int \dfrac{kv}{g + kv}\,\mathrm{d}v = -\int k\,\mathrm{d}x.$

Put $\qquad\qquad\qquad\qquad \dfrac{kv}{g + kv} = 1 - \dfrac{g}{g + kv}.$

Initially, $x = 0$, $v = u$; at highest point $x = h$, $v = 0$.

$$\int_u^0 \left(1 - \dfrac{g}{g + kv}\right)\mathrm{d}v = \int_0^h -k\,\mathrm{d}x,$$

$$\left[v - \dfrac{g}{k}\ln(g + kv)\right]_u^0 = \left[-kx\right]_0^h,$$

$$0 - \dfrac{g}{k}\ln(g) - u + \dfrac{g}{k}\ln(g + ku) = -kh.$$

Thus $\quad h = \dfrac{1}{k^2}\left[-g\ln\left(\dfrac{g + ku}{g}\right) + ku\right] = \dfrac{1}{k^2}\left[ku - g\ln\left(1 + \dfrac{ku}{g}\right)\right]$

(b) Using $\dfrac{\mathrm{d}v}{\mathrm{d}t} = -(g + kv) \;\Rightarrow\; \displaystyle\int_u^v \dfrac{\mathrm{d}v}{g + kv} = -\int_0^t \mathrm{d}t$

$$\left[\dfrac{1}{k}\ln(g + kv)\right]_u^v = \left[-t\right]_0^t$$

$$\dfrac{1}{k}\ln(g + kv) - \dfrac{1}{k}\ln(g + ku) = -t$$

$$\ln\left(\dfrac{g + kv}{g + ku}\right) = -kt. \qquad\qquad (1)$$

Antilogging, $\qquad g + kv = (g + ku)\,\mathrm{e}^{-kt},$

$$v = \dfrac{1}{k}\left[(g + ku)\,\mathrm{e}^{-kt} - g\right].$$

(c) The time taken to reach the highest point is found by putting $v = 0$ in equation (1):

$$t = -\dfrac{1}{k}\ln\left(\dfrac{g}{g + ku}\right) = \dfrac{1}{k}\ln\left(\dfrac{g + ku}{g}\right) = \dfrac{1}{k}\ln\left(1 + \dfrac{ku}{g}\right).$$

9.7 A particle moving on a straight line with speed v experiences a retardation of magnitude $b\mathrm{e}^{v/u}$, where b and u are constants. Given that the particle is travelling with speed u at time $t = 0$, show that the time t_1 for the speed to decrease to $u/2$ is given by

$$bt_1 = u\,(\mathrm{e}^{-1/2} - \mathrm{e}^{-1}).$$

Find the further time t_2 for the particle to come to rest.
Deduce that $t_2/t_1 = \mathrm{e}^{1/2}$.

Find, in terms of b and u, an expression for the distance travelled in decelerating from speed u to rest. (L)

- Retardation $= be^{v/u}$ so $\dfrac{dv}{dt} = -be^{v/u}$;

subject to $v = u$ when $t = 0$; $v = \dfrac{u}{2}$ when $t = t_1$.

$$\int_u^{u/2} -e^{-v/u}\, dv = \int_0^{t_1} b\, dt \quad \Rightarrow \quad \left[ue^{-v/u}\right]_u^{u/2} = \left[bt\right]_0^{t_1}.$$

$ue^{-1/2} - ue^{-1} = bt_1$,

thus $\qquad bt_1 = u\,(e^{-1/2} - e^{-1})$. $\hfill (1)$

Time taken to decrease speed to $\dfrac{u}{2}$ is given by $bt_1 = u\,(e^{-1/2} - e^{-1})$.

Changing the upper limits to $v = 0$, and $t = t_1 + t_2$, gives:

$$u\,(e^0 - e^{-1}) = b\,(t_2 + t_1)$$

$$b\,(t_2 + t_1) = u\,(1 - e^{-1}). \hfill (2)$$

$(2) - (1)$ gives $\qquad t_2 = \dfrac{u}{b}(1 - e^{-1/2})$

$$\frac{t_2}{t_1} = \frac{1 - e^{-1/2}}{e^{-1/2} - e^{-1}} = \frac{(1 - e^{-1/2})}{e^{-1/2}\,(1 - e^{-1/2})} = e^{1/2}.$$

Retardation $= -v\,\dfrac{dv}{dx} = be^{v/u}$.

Separating and integrating, $\qquad \displaystyle\int_u^0 -ve^{-v/u}\, dv = \int_0^{x_1} b\, dx,$

$$bx_1 = \left[vue^{-v/u}\right]_u^0 - \int_u^0 ue^{-v/u}\, dv = \left[vue^{-v/u}\right]_u^0 + \left[u^2 e^{-v/u}\right]_u^0$$

$$= u^2 - u^2 e^{-1} - u^2 e^{-1}.$$

Hence $x_1 = \dfrac{u^2}{b}(1 - 2e^{-1})$.

9.8 A new car of mass M experiences a resistance of kv^2 when moving in a straight line. The maximum speed attainable under a power P (assumed constant) is u. Find the distance travelled from rest in order to reach 4/5 of the maximum speed.

If the maximum force exerted by the brakes is R find the distance travelled in coming to an emergency stop from 4/5 of the maximum speed.

- Tractive force $= \dfrac{\text{power}}{\text{speed}} = \dfrac{P}{v}$,

accelerating force $= \dfrac{P}{v} - kv^2$.

At maximum speed $v = u$, acceleration $= 0$ so $k = \dfrac{P}{u^3}$,

thus acceleration $= \dfrac{dv}{dt} = v\,\dfrac{dv}{dx} = \dfrac{1}{M}\left(\dfrac{P}{v} - \dfrac{Pv^2}{u^3}\right)$

$$= \frac{P}{Mvu^3}(u^3 - v^3).$$

Now $v = 0$ when $x = 0$ and $v = \frac{4}{5}u$ when $x = X$ (say),

so, separating and integrating,

$$\int_0^{4u/5} \frac{v^2}{u^3 - v^3}\, dv = \int_0^X \frac{P}{Mu^3}\, dx,$$

$$-\frac{1}{3}\left[\ln(u^3 - v^3)\right]_0^{4u/5} = \left[\frac{Px}{Mu^3}\right]_0^X = \frac{PX}{Mu^3}.$$

$$X = \frac{Mu^3}{3P}\left[\ln(u^3) - \ln\left(u^3 - \frac{64}{125}u^3\right)\right] = \frac{Mu^3}{3P}\ln\frac{125}{61}.$$

Distance travelled while accelerating $= \dfrac{Mu^3}{3P}\ln\dfrac{125}{61}.$

In an emergency stop, the power is cut off and the brakes applied.

Accelerating force $= -\left(\dfrac{Pv^2}{u^3} + R\right)$, $\quad v\dfrac{dv}{dx} = -\dfrac{1}{Mu^3}(Pv^2 + Ru^3).$

Separate and integrate:

$$-\int_{4u/5}^0 \frac{v}{Pv^2 + Ru^3}\, dv = -\int_0^{X_1} \frac{1}{Mu^3}\, dx,$$

where X_1 is the stopping distance;

$$-\left[\frac{1}{2P}\ln(Pv^2 + Ru^3)\right]_{4u/5}^0 = \frac{X_1}{Mu^3}.$$

$$X_1 = \frac{Mu^3}{2P}\ln\frac{16Pu^2 + 25Ru^3}{25Ru^3} = \frac{Mu^3}{2P}\ln\left(1 + \frac{16P}{25Ru}\right).$$

Stopping distance $= \dfrac{Mu^3}{2P}\ln\left(1 + \dfrac{16P}{25Ru}\right).$

9.9 A particle is moving vertically downwards in a medium which exerts a resistance to the motion proportional to the speed of the particle. It is released from rest at O at zero time, and at time t its speed is v and its position is at a distance z below O. If the terminal velocity is U, prove that

(a) $\dfrac{dv}{dt} = g\left(1 - \dfrac{v}{U}\right),$ (b) $gz + Uv = Ugt.$ (OLE)

● Let the resistance of the medium be mkv.

Accelerating force $= ma = mg - mkv$,
$$a = g - kv.$$

When $a = 0$, $v = U$, so $g - kU = 0 \quad\Rightarrow\quad k = \dfrac{g}{U}.$

$$\frac{dv}{dt} = g - g\left(\frac{v}{U}\right) = g\left(1 - \frac{v}{U}\right). \tag{a}$$

$$\frac{dv}{dt} = g\frac{(U - v)}{U}.$$

128

Separating and integrating,

$$\int_0^v \frac{dv}{U-v} = \int_0^t \frac{g}{U}\,dt,$$

$$\left[-\ln(U-v)\right]_0^v = \left[\frac{g}{U}\,t\right]_0^t \quad \Rightarrow \quad \frac{g}{U}\,t = \ln\left(\frac{U}{U-v}\right) \qquad (1)$$

Alternatively, $\quad v\dfrac{dv}{dz} = \dfrac{g}{U}(U-v)$

$$\int \frac{v}{U-v}\,dv = \int \frac{g}{U}\,dz.$$

Putting $\dfrac{v}{U-v} = -1 + \dfrac{U}{U-v}$ gives

$$\int_0^v \left(-1 + \frac{U}{U-v}\right)dv = \int_0^z \frac{g}{U}\,dz,$$

$$\left[-v - U\ln(U-v)\right]_0^v = \left[\frac{gz}{U}\right]_0^z,$$

$$-v + U\ln\left(\frac{U}{U-v}\right) = \frac{g}{U}\,z,$$

$$-v + U\,\frac{g}{U}\,t = \frac{g}{U}\,z \qquad \text{from (1)}$$

$$-Uv + Ugt = gz;$$

i.e. $\qquad\qquad\qquad\qquad Ugt = gz + Uv. \qquad\qquad\qquad$ (b)

9.10 A particle P moves along a straight line such that, when its speed is v m s^{-1}, its retardation is $4v^{n+1}$ m s^{-2}, where $n\;(> -1)$ is a constant. The speed of P at time $t = 0$ s is u m s^{-1}.
(a) Show that, for $n = 0$, $v = ue^{-4t}$.
(b) Find similarly an expression for v, in terms of u, n and t, when $n \neq 0$.
(c) When $n = 3$ obtain an expression for the speed with which P is moving when it has travelled a distance of s m from its initial position. (AEB 1984)

• Retardation $= 4v^{n+1}$, $\quad \Rightarrow \quad \dfrac{dv}{dt} = -4v^{n+1};$

$$-\int_u^v \frac{dv}{v^{n+1}} = 4\int_0^t dt.$$

(a) When $n = 0$, $\qquad \left[-\ln v\right]_u^v = \left[4t\right]_0^t.$

$$\ln\frac{u}{v} = 4t; \quad u = ve^{4t}, \quad v = ue^{-4t}.$$

(b) When $n \neq 0$, $\left[\dfrac{1}{nv^n}\right]_u^v = \left[4t\right]_0^t$.

i.e. $\dfrac{1}{v^n} - \dfrac{1}{u^n} = 4nt, \quad \Rightarrow \quad \dfrac{1}{v^n} = 4nt + \dfrac{1}{u^n}$

$$\Rightarrow \quad v^n = \frac{1}{4nt + 1/u^n},$$

$$v^n = \frac{u^n}{4ntu^n + 1}, \quad v = \frac{u}{(4ntu^n + 1)^{1/n}}.$$

(c) When $n = 3$, $v\dfrac{dv}{ds} = -4v^4$,

$$\int_u^v -\frac{dv}{v^3} = \int_0^s 4\,ds \quad \Rightarrow \quad \left[\frac{1}{2v^2}\right]_u^v = \left[4s\right]_0^s$$

$$\frac{1}{2v^2} - \frac{1}{2u^2} = 4s \quad \Rightarrow \quad \frac{1}{2v^2} = 4s + \frac{1}{2u^2}$$

$$2v^2 = \frac{1}{4s + 1/2u^2}$$

$$v^2 = \frac{2u^2}{2(8u^2s + 1)}$$

$$v = \left(\frac{u^2}{8u^2s + 1}\right)^{1/2}.$$

9.3 Exercises

9.1 A particle moves in a straight line with variable acceleration $\dfrac{k}{1 + v}$ m s^{-2}, where k is a constant and v m s^{-1} is the speed of the particle when it has travelled a distance x m. Find the distance moved by the particle as its speed increases from 0 to u m s^{-1}.

(L)

9.2 A particle moving along a straight line has a velocity v when a distance x from a fixed point O on the line. Show that the acceleration of the particle may be expressed as $v\dfrac{dv}{dx}$ or $\dfrac{d}{dx}\left(\dfrac{1}{2}v^2\right)$.

The force of attraction experienced by a mass m at a distance x $(> a)$ from the centre O of the earth towards O is $\dfrac{mga^2}{x^2}$, where a is the radius of the earth and g is a constant.

A particle mass m starts from the surface of the earth with a velocity u directly away from O. Find the subsequent velocity v, when the particle is distance x from O, in terms of u, g, a and x.

Deduce that if $u^2 > 2ga$, the particle will escape from the earth's attraction.

9.3 In this question take g as 10 m s^{-2}. The resistance to motion of a car whose mass is 750 kg is proportional to its speed. When climbing a slope of $\sin^{-1}$ (1/25) at a constant speed of 10 m s^{-1}, the engine works at a rate of 30 kW. When the car is travelling down the same slope with the engine still working at a constant rate of 30 kW, it has speed v m s^{-1} at time t seconds after starting its descent. Show that

$$25\frac{dv}{dt} = \frac{(10 + v)(100 - 9v)}{v}.$$

Find, correct to 2 significant figures, the time for the car to reach a speed of 10 m s^{-1} from rest on its path down the slope. (SUJB)

9.4 One method of dyeing material is to immerse the cloth in a bath of water to which has been added d grams of concentrated dye. The material absorbs the dye at a rate equal to half the amount of dye remaining.

If $\frac{3}{4}d$ grams of dye need to be absorbed to reach the desired colour, find the time taken to reach completion.

An alternative process is to keep the amount of dye present in the water constant at d grams by continuously adding dye throughout the process.

Find the time now taken to complete the process.

9.5 The stopping distance, x, is the distance travelled by a vehicle in the interval between the driver seeing an obstacle and the vehicle coming to rest. It is calculated on the assumptions that:

a time interval T elapses between the obstacle being seen and the brakes being applied, during which time there is zero acceleration;

the brakes, when applied, immediately produce a retardation equal to that produced when the vehicle slides along the road. The coefficient of friction between the road and the tyres is a constant value μ.

Given that the initial speed is $u = 20$ m s^{-1}, $T = 2$ s, $\mu = 0.3$ and $g = 10$ m s^{-2}, find x in metres.

A more realistic model of the braking effect is that the retardation is dependent on the speed of the vehicle.

With the above values of u and T, and given that the retardation at speed v m s^{-1} is $\dfrac{660}{(v + 200)}$ m s^{-2}, calculate the new value of x in metres.

(AEB 1983)

9.6 A ball of mass m is thrown with initial velocity $\mathbf{u}$ and experiences a resistance to motion of $\frac{1}{2}m\,|\mathbf{v}|$ where $\mathbf{v}$ is the velocity at any instant of time t.

Find a differential equation giving $\dfrac{d\mathbf{v}}{dt}$ in terms of $\mathbf{v}$ and $\mathbf{g}$.

Hence find an expression for T, the time taken to reach maximum height.

9.7 A particle is projected vertically upwards with speed U in a medium where there is a resistance kv^2 per unit mass when v is the speed. Prove that the maximum height reached is $\dfrac{1}{2k}\ln\left(1 + \dfrac{kU^2}{g}\right)$, and find the time taken to reach this height.

(OLE)

9.8 A car of mass M kg has a power of $16M$ watts and experiences a resistance of Mv where v m s^{-1} is the velocity at time t, and x is the displacement.

When $t = 0$, $x = 0$ and $v = 2$.

Find

(a) an expression for $\dfrac{\mathrm{d}v}{\mathrm{d}t}$ and hence v in terms of t,

(b) the maximum speed,

(c) x in terms of v.

9.9 The resultant force acting on a train of mass m starting from rest on a level track is a constant P for speeds less than V. For speeds greater than V the power exerted by the resultant force has a constant value PV. Find the time taken to reach a speed $2V$ from rest, and the corresponding distance travelled. (L)

9.4 Brief Solutions to Exercises

9.1 $\quad v\,\dfrac{\mathrm{d}v}{\mathrm{d}x} = \dfrac{k}{1+v} \quad \Rightarrow \quad \displaystyle\int v\,(1+v)\,\mathrm{d}v = \int k\,\mathrm{d}x .$

When $x = 0$, $v = 0$. Let $x = x_1$ when $v = u$; then $x_1 = \dfrac{u^2}{6k}\,(3 + 2u)$.

9.2 Bookwork: see Fact Sheet, section 9.1(a).

$$v\,\frac{\mathrm{d}v}{\mathrm{d}x} = -\frac{ga^2}{x^2} \quad \Rightarrow \quad \int_u^v v\,\mathrm{d}v = -ga^2\int_a^x \frac{1}{x^2}\,\mathrm{d}x ,$$

$$v^2 - u^2 = 2ga^2\left(\frac{1}{x} - \frac{1}{a}\right) \quad \Rightarrow \quad v = u^2 - \left[2ga\,\frac{(x-a)}{x}\right]^{1/2} .$$

Particle will escape if $v > 0$ for all x

$$\Rightarrow \quad u^2 > 2ga\,\frac{(x-a)}{x} = 2ga - \frac{2ga^2}{x} \quad \text{for all } x > a,$$

i.e. if $u^2 > 2ga$.

9.3 Moving uphill. At constant speed the sum of forces is zero.

$k = 270 \quad \Rightarrow \quad$ resistance $= 270v$.

Moving downhill,

$$\text{accel.} = \frac{1}{750}\left(300 + \frac{30\,000}{v} - 270v\right)$$

$$\Rightarrow \quad 25\,\frac{\mathrm{d}v}{\mathrm{d}t} = \frac{(10+v)(100-9v)}{v} .$$

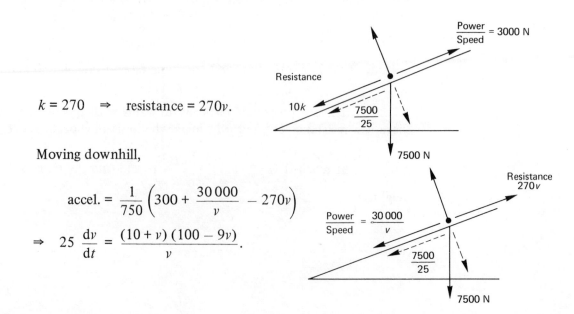

Limits of v, 0 to 10; of t, 0 to T.

$$\int_0^{10} \frac{25v}{(10+v)(100-9v)} \, dv = \int_0^T dt.$$

Use partial fractions to get $T = \dfrac{25}{19} \left[-\dfrac{10}{9} \ln(100-9v) - \ln(10+v) \right]_0^{10}.$

$$T = \frac{25}{19}\left[\frac{10}{9} \ln 10 - \ln 2 \right] = 2.5 \text{ s.}$$

9.4 Let x be the amount of dye absorbed by the cloth.

(a) $\dfrac{dx}{dt} = \dfrac{(d-x)}{2}, \qquad \displaystyle\int_0^{3d/4} \dfrac{dx}{d-x} = \int_0^{T_1} \dfrac{1}{2} \, dt.$

$T_1 = 4 \ln 2 = 2.77.$

(b) $\dfrac{dx}{dt} = \dfrac{d}{2}, \qquad T_2 = \dfrac{3}{2} = 1.5.$

9.5 (a) Thinking distance = 40 m,   accel. = -3 m s^{-2}.

Total distance $= 40 + \dfrac{400}{6} = 106.7$ m.

(b) Accel. $= \dfrac{-660}{(v+200)} = v \dfrac{dv}{dx}.$

$$\int_{20}^0 -v\,(v+200) \, dv = \int_0^{x_1} 660 \, dx.$$

Total distance $= 40 + 64.6 = 104.6$ m.

9.6 Resistance $= + \dfrac{1}{2} m \,|\mathbf{v}| \quad \Rightarrow \quad \dot{\mathbf{v}} = \mathbf{g} - \dfrac{1}{2} \mathbf{v}.$

For $\dot{\mathbf{v}} + \dfrac{1}{2} \mathbf{v} = \mathbf{g}$, use an integrating factor $e^{t/2}$, $\Rightarrow \quad e^{t/2} \mathbf{v} = 2\mathbf{g}e^{t/2} + \mathbf{c}$,

giving $\mathbf{v} = 2\mathbf{g}(1 - e^{-t/2}) + \mathbf{u}e^{-t/2}.$

At max. height $\mathbf{v}$ is horizontal, $\mathbf{v} \cdot \mathbf{g} = 0$, giving $T = 2 \ln \left(1 - \dfrac{\mathbf{u} \cdot \mathbf{g}}{2g^2} \right).$

9.7 $v \dfrac{dv}{dx} = -(kv^2 + g) \quad \Rightarrow \quad \displaystyle\int_U^v \dfrac{v\,dv}{kv^2 + g} = \int_0^x -dx$

$$x = \dfrac{1}{2k} \ln\left(\dfrac{kU^2 + g}{kv^2 + g}\right).$$

At maximum height $v = 0$. Maximum height $= \dfrac{1}{2k} \ln\left(1 + \dfrac{kU^2}{g}\right).$

$$\dfrac{dv}{dt} = -(kv^2 + g) \quad \Rightarrow \quad \int_U^0 \dfrac{dv}{(v^2 + g/k)} = -\int_0^T k\,dt,$$

where T is the time taken to reach the maximum height.

$$\left[\sqrt{\left(\dfrac{k}{g}\right)} \tan^{-1} v \sqrt{\left(\dfrac{k}{g}\right)}\right]_U^0 = -kT \quad \Rightarrow \quad T = \dfrac{1}{\sqrt{(kg)}} \tan^{-1}\left[U \sqrt{\left(\dfrac{k}{g}\right)}\right].$$

9.8 (a) $\dfrac{dv}{dt} = \dfrac{16}{v} - v \quad \Rightarrow \quad \displaystyle\int_2^v \dfrac{v}{16 - v^2}\,dv = \int_0^t dt,$

$$t = \dfrac{1}{2} \ln\left(\dfrac{12}{16 - v^2}\right) \quad \Rightarrow \quad v = \sqrt{(16 - 12e^{-2t})}.$$

(b) As $t \to \infty$, $v \to 4$. Maximum speed $= 4$ m s^{-1}.

(c) $\quad v \dfrac{dv}{dx} = \dfrac{16 - v^2}{v} \quad \Rightarrow \quad \displaystyle\int_2^v \dfrac{v^2}{16 - v^2}\,dv = \int_0^x dx,$

$$\dfrac{v^2}{16 - v^2} = -1 + \dfrac{2}{4 - v} + \dfrac{2}{4 + v}.$$

$$x = \left[-v + 2\ln\left(\dfrac{4 + v}{4 - v}\right)\right]_2^v,$$

$$x = 2 - v + 2\ln\left(\dfrac{(4 + v)}{3(4 - v)}\right).$$

9.9 When vel. $v < V$, force $= P$; $\quad a = \dfrac{P}{m}$ (a constant).

Time to reach V is $\dfrac{Vm}{P}$.

Distance travelled $= \dfrac{mV^2}{2P}$.

When vel. $v > V$, power $= PV$, accelerating force $= \dfrac{PV}{v} \quad \Rightarrow \quad \dfrac{dv}{dt} = \dfrac{PV}{mv},$

$$\int_V^{2V} mv\,dv = \int_{Vm/P}^{T_1} PV\,dt \qquad \text{where } T_1 \text{ is the total time.}$$

$\dfrac{3}{2} V^2 m = PVT_1 - mV^2$, total time taken is $T_1 = \dfrac{5mV}{2P}$.

$$a = v\,\frac{\mathrm{d}v}{\mathrm{d}x} = \frac{PV}{mv} \quad \Rightarrow \quad \int_{V}^{2V} mv^2\,\mathrm{d}v = PV \int_{m\,V^2/2P}^{X} \mathrm{d}x$$

where X is the total distance travelled.

$\dfrac{7mV^3}{3} = PVX - \dfrac{mV^3}{2}$, total distance travelled is $X = \dfrac{17mV^2}{6P}$.

10 Work. Energy. Simple Harmonic Motion. Elastic Strings

Kinetic energy and work done.
Hooke's law and the work done in stretching an elastic string or spring.
Elastic and gravitational potential energy.
Applications of the principle of conservation of energy to problems involving elastic and gravitational forces.
The general solution of $\ddot{x} = -\omega^2 x$.

10.1 Fact Sheet

(a) Hooke's Law

When an elastic string, natural length a, modulus of elasticity λ, is stretched by a distance x, the tension in the string is $T = \lambda \dfrac{x}{a}$.

For a spring in compression, the force is a thrust.

(b) Work Done

The work done by a force $\mathbf{F}$ moving from A to B is $\displaystyle\int_A^B \mathbf{F} \,.\, \mathrm{d}\mathbf{r}$.

For motion in a straight line, work done $= \displaystyle\int_A^B F \,\mathrm{d}x$.

$F = ma = mv \dfrac{\mathrm{d}v}{\mathrm{d}x}$, so work done $= \displaystyle\int_{v_1}^{v_2} mv \,\mathrm{d}v = \tfrac{1}{2}mv_2^2 - \tfrac{1}{2}mv_1^2$

$= $ increase in kinetic energy.

For motion under gravity, work done $= \displaystyle\int_{h_1}^{h_2} mg\,dx = mgh_2 - mgh_1$

= increase in gravitational potential energy.

For stretching an elastic string, work done $= \displaystyle\int_{x_1}^{x_2} \dfrac{\lambda x}{a}\,dx = \dfrac{\lambda(x_2^2 - x_1^2)}{2a}$

= increase in elastic potential energy.

(c) Conservation of Energy

When a particle is oscillating on the end of an elastic string, the sum of
(i)   kinetic energy (k.e.),
(ii) gravitational potential energy (g.p.e.),
(iii) elastic potential energy (e.p.e),
remains constant.

(d) Simple Harmonic Motion

$\ddot{x} = -\omega^2 x$ has a general solution. $x = A\sin\omega t + B\cos\omega t$, period $\dfrac{2\pi}{\omega}$.

Note: ωt is measured in radians.
(i)   If $t = 0$ when the particle passes through the centre of oscillation ($x = 0$), then $x = A\sin\omega t$, amplitude A.
(ii) If $t = 0$ when the particle has maximum displacement ($x = B$), then $x = B\cos\omega t$, amplitude B.
(iii) In other cases, $x = A\sin\omega t + B\cos\omega t$, or $x = C\cos(\omega t + \alpha)$, amplitude $\sqrt{(A^2 + B^2)} = C$.

Speed as a Function of Displacement

$\ddot{x} = -\omega^2 x \quad\Rightarrow\quad v\dfrac{dv}{dx} = -\omega^2 x \quad\Rightarrow\quad v = \omega\sqrt{(a^2 - x^2)}.$

Notes
(i)   Maximum displacement $x = \pm a$ occurs when $v = 0$.
(ii) Maximum speed $v = a\omega$ occurs when $x = 0$.
(iii) Maximum acceleration $\ddot{x} = \mp\omega^2 a$ occurs when $x = \pm a$.
(iv) The accelerating force always acts towards the centre of oscillation.
(v)   For a particle on an elastic string the centre of oscillation is the position of static equilibrium.

(e) External Forces

For a particle moving against external applied forces (such as friction and weight),

the decrease in the kinetic energy = the work done against the external forces.

10.2 Worked Examples

10.1 A mass of 6 kg is accelerated vertically upwards from a position of rest. When it attains a height of 7 m it has a velocity of 8 m s^{-1}. Neglecting air resistance, find the work done by the force causing the motion. (Assume $g = 10$ m s^{-2}.)

- Work done by force raising 6 kg through 7 m = $(6g)(7) = 420$ J.
  Work done giving 6 kg a velocity of 8 m s^{-1} = $\frac{1}{2}(6)(8)^2 = 192$ J.
  Total work done by the force = 612 J.

10.2 A particle P has a position vector relative to a fixed origin O of $\mathbf{r} = 1\mathbf{i} + 2t\mathbf{j} + 3t^2\mathbf{k}$, where t is the time.

A force $\mathbf{F}$ is exerted on the particle where $\mathbf{F} = \dfrac{-3\mathbf{r}}{|\mathbf{r}|}$.

The work done by a force $\mathbf{F}$ displaced by $\delta\mathbf{r}$ is defined vectorially $\mathbf{F} \cdot \delta\mathbf{r}$.

Show that the work done by $\mathbf{F}$ in the time interval between $t = 1$ and $t = 2$ may be expressed as follows:

$$-3 \int_1^2 \frac{18t^3 + 4t}{(9t^4 + 4t^2 + 1)^{1/2}}\, dt.$$

Find the work done in this time interval.

- $\mathbf{r} = 1\mathbf{i} + 2t\mathbf{j} + 3t^2\mathbf{k}, \quad \mathbf{F} = \dfrac{-3(1\mathbf{i} + 2t\mathbf{j} + 3t^2\mathbf{k})}{\sqrt{(1 + 4t^2 + 9t^4)}}.$

$$\frac{d\mathbf{r}}{dt} = 0\mathbf{i} + 2\mathbf{j} + 6t\mathbf{k}.$$

Work done in time t $\qquad = \displaystyle\int \mathbf{F} \cdot \frac{d\mathbf{r}}{dt}\, dt.$

$$\mathbf{F} \cdot \frac{d\mathbf{r}}{dt} = \frac{-3(4t + 18t^3)}{\sqrt{(1 + 4t^2 + 9t^4)}}.$$

Work done between $t = 1$ and $t = 2$ $= -3 \displaystyle\int_1^2 \frac{18t^3 + 4t}{\sqrt{(9t^4 + 4t^2 + 1)}}\, dt$

$$= -3 \left[(9t^4 + 4t^2 + 1)^{1/2} \right]_1^2$$

$$= +3\,(\sqrt{14} - \sqrt{161})$$

$$= -26.8.$$

10.3 Show that the work done in stretching an elastic string of natural length l

and modulus of elasticity λ from length $(l + x_1)$ to $(l + x_2)$ is $\dfrac{\lambda(x_2^2 - x_1^2)}{2l}$.

A string has a natural length of 1 m. If the work done in stretching the string from a length of 1.1 m to 1.2 m is 0.3 J, find the energy stored in the string when it has a length of 1.3 m.

- Work done stretching string from $(l + x_1)$ to $(l + x_2)$

$$= \int_{x_1}^{x_2} T\, dx = \int_{x_1}^{x_2} \frac{\lambda x}{l}\, dx = \frac{\lambda}{2l}(x_2^2 - x_1^2).$$

Natural length = 1   so $0.3 = \frac{\lambda}{2}[(0.2)^2 - (0.1)^2] = \frac{\lambda}{2}(0.03).$

Hence $\lambda = \dfrac{0.6}{0.03} = 20.$

When $x = 0.3$, stored energy $= \dfrac{20}{2}(0.3)^2 = 10(0.09)$

$$= 0.9 \text{ J}.$$

10.4 One end of a light elastic string of natural length l and modulus of elasticity $3mg$ is attached to a fixed point O. To the other end is attached a particle P of mass m. The particle is projected vertically downwards from O with a speed $\sqrt{(3lg)}$. By the principle of conservation of energy, or otherwise, find the speed of P when it is at a depth x, $(x > l)$, below O. Show that the greatest depth below O reached by P is $\frac{8}{3}l$.

Find the maximum speed of P.

● Take the potential energy datum as O.

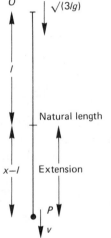

When at O, kinetic energy $= \frac{1}{2}m(3lg) = \dfrac{3mgl}{2}$,

potential energy of particle $= 0$,
potential energy in string $\quad = 0$.

When particle has descended a distance x $(x > l)$, extending the string by $(x - l)$,
k.e. of particle $= \frac{1}{2}mv^2$, p.e. of particle $= -mgx$,

$$\text{p.e. in string} = \frac{\lambda\,(\text{ext.})^2}{2l} = \frac{3mg(x - l)^2}{2l}.$$

By conservation of energy,

$$\frac{3}{2}mgl = \frac{1}{2}mv^2 - mgx + \frac{3mg(x - l)^2}{2l},$$

$$v^2 = 3gl + 2gx - \frac{3g(x - l)^2}{l}$$

$$= \frac{g}{l}(3l^2 + 2xl - 3x^2 + 6xl - 3l^2),$$

$$v^2 = \frac{g}{l}(8xl - 3x^2). \tag{1}$$

Speed of P when at a depth x $(x > l)$ below O is $v = \sqrt{\left(\dfrac{g}{l}(8xl - 3x^2)\right)}.$

$v = 0$ when $x = 0$ (not applicable), or $x = \dfrac{8}{3}l.$

Greatest depth below O reached by P is $\dfrac{8}{3}l.$

Maximum speed is attained when $\dfrac{\mathrm{d}v}{\mathrm{d}t} = v\dfrac{\mathrm{d}v}{\mathrm{d}x} = 0.$

Differentiating (1), $2v\dfrac{\mathrm{d}v}{\mathrm{d}x} = \dfrac{g}{l}(8l - 6x),$

Maximum speed occurs when $x = \frac{4}{3}l.$

Then $v^2 = \dfrac{g}{l}\left(\dfrac{32l^2}{3} - \dfrac{48l^2}{9}\right) = \dfrac{16gl}{3}$.

Maximum speed of P is $4\sqrt{\left(\dfrac{gl}{3}\right)}$.

10.5 (multiple choice) A particle P moves along Ox with simple harmonic motion, O being the centre of the motion. At time t, the displacement of P from O is x.

(1) $x = b\cos\omega t$ where b and ω are non-zero constants.

(2) The period is $\dfrac{2\pi}{\omega}$.

Determine if: A 1 $\Rightarrow$ 2, 2 $\not\Rightarrow$ 1.
       B 2 $\Rightarrow$ 1, 1 $\not\Rightarrow$ 2.
       C 1 $\iff$ 2.
       D 1 denies 2, 2 denies 1.
       E None of these.   (L)

- If $x = b\cos\omega t$ the period $= \dfrac{2\pi}{\omega}$, i.e. 1 $\Rightarrow$ 2.

 If $T = \dfrac{2\pi}{\omega}$ then general s.h.m. solution is $x = A\cos\omega t + B\sin\omega t$, i.e. 2 $\not\Rightarrow$ 1.

<u>Answer A</u>

10.6 Owing to the tides the water height in an estuary may be assumed to rise and fall with time in simple harmonic motion. At a certain place a danger of flooding exists when the water height is above 1.5 m. One day the high tide had a height of 1.6 m at 01.00 and the following low tide a height of 0.4 m at 07.30. Find the times that day when there was a danger of flooding, assuming that the following high tide was also 1.6 m.

- Let d metres be the depth of water.
  Mean depth of water = 1.0 m, so $d = 1.0 + 0.6\cos\omega t$, where $t = 0$ at 01.00.

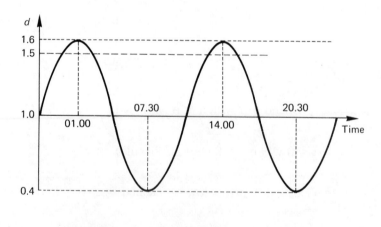

Period = 13 hours = $\dfrac{2\pi}{\omega}$ so $\omega = \dfrac{2\pi}{13}$.

Danger exists when $d > 1.5$.

Then $\cos\dfrac{2\pi}{13}t > \dfrac{5}{6}$ $\Rightarrow$ $\left|\dfrac{2\pi}{13}t\right| < 0.5856$.

140

$|t| \leqslant 1.21$ hours = 1 hour 12.6 minutes.

There was a danger of flooding for 1 hour 13 minutes on either side of high tide.

The danger existed from midnight (in fact from 23.47 the previous day) to 02.13 and from 12.47 to 15.13.

10.7 A light elastic string with natural length a and modulus of elasticity mg, has a particle of mass m attached to one end. The other end is attached to the fixed point O, and the particle hangs in equilibrium. A particle of mass M is then attached to m. Prove that the subsequent motion is simple harmonic with amplitude aM/m, and find the period.

When the string is stretched to its fullest extent the particle M falls off and in the subsequent motion the particle just reaches O. By considering potential energy, or otherwise, show that $M = \frac{1}{2}\sqrt{3}m$.　　　　　　　　　　　(OLE)

- $T = \dfrac{\lambda\,(\text{ext.})}{\text{nat. length}}$,　$\lambda = mg$,　nat. length $= a$.

With particle mass m attached.

When in equilibrium, $T = mg = \dfrac{mg\,(\text{ext.})}{a}$　　so extension is a.

With a total mass $m + M$ attached.

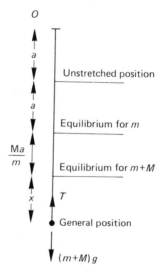

When in equilibrium, $(m + M)g = \dfrac{mg\,(\text{ext.})}{a}$　so extension is now $\dfrac{(m + M)\,a}{m}$.

With a further extension x,　$T = \dfrac{mg\,[(m + M)\,a/m + x]}{a}$.

The equation of motion is

$$(m + M)\,\ddot{x} = (m + M)g - T$$

$$= (m + M)g - \frac{mg\,[(m + M)\,a/m + x]}{a}$$

$$\Rightarrow\quad \ddot{x} = -\frac{mg}{(m + M)\,a}\,x = -\omega^2 x \qquad \text{where} \qquad \omega^2 = \frac{mg}{(m + M)\,a}.$$

141

This is the simple harmonic motion equation (s.h.m.) with period

$$\frac{2\pi}{\omega} = 2\pi\sqrt{\left(\frac{(m+M)a}{mg}\right)}$$

and general solution $x = A \cos \omega t + B \sin \omega t$.

When $t = 0$, $x = -\dfrac{Ma}{m}$ (the equilibrium position for m)

and $\dot{x} = 0$ (since it starts from rest).

Thus $A = -\dfrac{Ma}{m}$ and $B = 0$, giving the particular solution

$$x = -\frac{Ma}{m} \cos \omega t.$$

Amplitude of motion $= \dfrac{Ma}{m}$, period $= 2\pi\sqrt{\left(\dfrac{(m+M)a}{mg}\right)}.$

At lowest point, extension $= \dfrac{(m+M)a}{m} + \dfrac{Ma}{m} = \dfrac{(m+2M)a}{m}.$

Potential energy of the string $= \dfrac{mg\,[(m+2M)\,a/m]^2}{2a}.$

When the particle m just reaches O the string has lost all its potential energy since it is now slack, but the particle m has gained potential energy

$$mg\left(\frac{(m+2M)}{m}a + a\right).$$

In both positions the kinetic energy $= 0$.
Equating energies,

$$\frac{mg\,(m+2M)^2\,a^2}{2am^2} = \frac{mg\,(2m+2M)\,a}{m}$$

$$(m+2M)^2 = 2m\,(2m+2M)$$

$$m^2 + 4mM + 4M^2 = 4m^2 + 4mM$$

$$4M^2 = 3m^2 \quad \Rightarrow \quad M = \tfrac{1}{2}\sqrt{3}m.$$

10.8   Show that small oscillations of a simple pendulum of length l are simple harmonic with period $2\pi\sqrt{(l/g)}$. A pendulum clock beats seconds (i.e. one half-period = 1 second) at a point where $g = 9.812$ m s^{-2}. Find the length of the pendulum correct to 3 significant figures. If the clock is moved to a place where $g = 9.921$ m s^{-2}, will the clock gain or lose? Find how much it will gain or lose during one day. To what length should the pendulum be altered if it is to register correctly?   (SUJB)

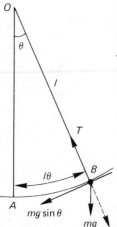

● Tangential component of force $= mg \sin\theta$ towards A, so
  tangential acceleration $= g\sin\theta \approx g\theta$ where θ is small and measured in radians.

  Arc length $AB = s = l\theta$, so $\theta = \dfrac{s}{l}$ and acceleration towards $A = \dfrac{gs}{l}$,

i.e. $-\ddot{s} = \dfrac{g}{l}s.$

This is simple harmonic motion with period $2\pi\sqrt{\left(\dfrac{l}{g}\right)}.$

For the clock, 1 period = 2 seconds, so $2\pi\sqrt{\left(\dfrac{l}{g}\right)} = 2 \quad \Rightarrow \quad l = \dfrac{g}{\pi^2}$ m.

When $g = 9.812$, $l = 99.4$ cm $= 0.994$ m.

When $g = 9.921$, period $= 2\pi \sqrt{\left(\dfrac{0.994}{9.921}\right)} = 1.989$ seconds.

This is less than 2 seconds so the clock will gain.

Number of 'beats' in 1 day $= \dfrac{24\,(60)\,(60)}{0.9945} = 86\,886$.

Number of seconds in 1 day $= 24\,(60\,(60)) = 86\,400$.

Thus clock gains 486 seconds $= 8$ min 6 s per day.

To register correctly, $l = \dfrac{9.921}{\pi^2} = 1.005$ m, i.e. it should be lengthened by 1.1 cm.

10.9 A light elastic string, of modulus $\lambda = 4mg$ and natural length a, has one end attached to a fixed point O. Two particles A and B of masses $3m$ and $2m$ are fastened to the other end and the system hangs in equilibrium. Show that the extension is $5a/4$.

One particle falls off. Find the different distances below O to which the remaining particle will rise. Show that in one case the remaining particle moves in complete cycles of simple harmonic motion and find the period of oscillation.

● When in equilibrium $T = 5mg = \dfrac{4mg\,(\text{ext.})}{a}$. Thus extension $= \dfrac{5a}{4}$.

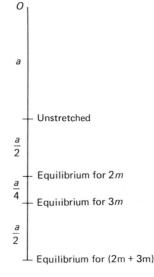

Case (a): $3m$ falls off, $2m$ remains attached.

Equilibrium position with $2m$ attached has an extension x

where $2mg = \dfrac{4mgx}{a}$ so $x = \dfrac{a}{2}$.

At lowest point extension is a further $\dfrac{3a}{4}$.

Thus the motion will commence as s.h.m. with amplitude $\dfrac{3a}{4}$.

However, $\dfrac{3a}{4} > \dfrac{a}{2}$, so string will go slack and motion will cease to be s.h.m.

By conservation of energy:

At lowest point, extension = $\dfrac{5a}{4}$, kinetic energy = 0,

potential energy stored in string = $\dfrac{4mg}{2a}\,(5a/4)^2 = \dfrac{25}{8}\,mga$.

At the highest point, distance h below O, the particle has gained potential energy $2mg\left(\dfrac{9a}{4}-h\right)$ and has zero kinetic energy, and the string has lost all its potential energy.

Thus $2mg\left(\dfrac{9a}{4}-h\right) = \dfrac{25}{8}\,mga, \quad \Rightarrow \quad h = \dfrac{11a}{16}$.

The highest point attained by mass $2m$ is $\dfrac{11a}{16}$ below O.

Case (b): $2m$ falls off, $3m$ remains attached.

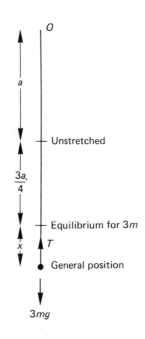

Equilibrium position for $3m$ has an extension $\dfrac{3a}{4}$.

At lowest point the extension is a further $\dfrac{a}{2}$.

Thus string will still be extended by $\dfrac{a}{4}$ when the particle mass $3m$ reaches its highest point, a distance $\dfrac{5a}{4}$ below O.

At a distance x below the $(3m)$ equilibrium position,

$$3m\ddot{x} = 3mg - T = 3mg - 4mg\,\frac{(3a/4 + x)}{a},$$

$$3m\ddot{x} = -\frac{4mg}{a}x \quad \text{or} \quad \ddot{x} = -\frac{4g}{3a}x = -\omega^2 x.$$

This is s.h.m. with period $2\pi\sqrt{\left(\dfrac{3a}{4g}\right)}$.

144

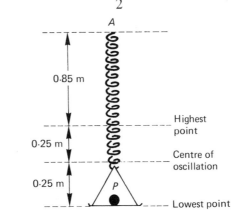

10.10 A scale pan is suspended from a fixed point A by a light elastic spring and a particle P of mass 0.3 kg is placed in the pan and attached to it with adhesive as shown.

The pan is pulled down from its equilibrium position and set in motion so that P moves in a vertical line through A with the base of the pan remaining horizontal.

Given that the motion of P is simple harmonic, with period $\frac{\pi}{5}$ s, and that the maximum and minimum distances of P below A are 1.35 m and 0.85 m respectively, find

(a) the distance below A of the centre O of the oscillation, the amplitude of the oscillation and the maximum speed obtained by P,

(b) the time to move directly upwards a distance of 0.125 m from O,

(c) the maximum force, normal to the scale pan, that the adhesive has to exert on P,

(d) the length of the spring, when, in the absence of adhesive, P would leave the pan. (Take g to be 10 m s^{-2}.)   (AEB 1984)

• (a) Centre of oscillation is $\dfrac{1.35 + 0.85}{2}$ m below A, i.e. 1.1 m below A.

Amplitude $a = \dfrac{1.35 - 0.85}{2} = 0.25$ m.

Period $\dfrac{2\pi}{\omega} = \dfrac{\pi}{5}$ $\Rightarrow$ $\omega = 10$.

Maximum speed = $\omega a = 2.5$ m s^{-1}.

(b) $x = 0.25 \sin 10t$, where x is the upward displacement from O, starting the timing at O.
When $x = 0.125$, $0.125 = 0.25 (\sin 10t)$,

$$\sin 10t = \frac{1}{2}, \quad \Rightarrow \quad 10t = \frac{\pi}{6}, \; t = \frac{\pi}{60} \text{ s.}$$

Particle takes 0.0524 s to move upwards a distance of 0.125 m from O.

(c) Maximum force that the glue has to exert is at the highest point, when a maximum acceleration downwards of $a\omega^2$ is experienced.

Accelerating force required = $0.3 (0.25) (100) = 7.5$ N.
Weight of particle = 3 N, so adhesive has to exert a force of 4.5 N.

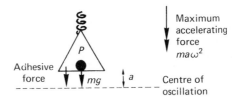

(d) If the particle is not glued it will leave the pan when $R = 0$, i.e. when the particle has an acceleration downwards of g,
i.e. when $-\omega^2 x = -10$;   but $\omega^2 = 100$, so $x = +0.1$.
$\Rightarrow$ length of spring $= 1$ m when P leaves the pan.

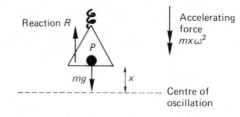

10.11   The top d metres of a slope, of inclination $30°$ to the horizontal, is covered with smooth ice, below which there is no ice and the surface is rough with coefficient of friction $\mu = \sqrt{3}$.

A toboggan starts at the top with an initial speed $\sqrt{(2gd)}$ m s^{-1}. By considering the work done against external forces, find the total distance travelled down the slope by the toboggan before coming to rest.

• Let the mass of the toboggan be m kg.

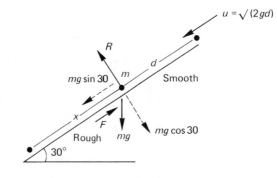

Normal reaction $R = mg \cos 30° = \dfrac{mg\sqrt{3}}{2}$,

Frictional force $F = \mu R = \dfrac{3mg}{2}$.

Total distance travelled down the slope $= d + x$.
Work done against external forces $=$ work done against friction $-$ loss of p.e.
$$= Fx - mg \sin 30° \, (d + x)$$

$$= \frac{3mgx}{2} - \frac{mg}{2} (d + x).$$

Kinetic energy lost by toboggan $= \frac{1}{2} m \, (2gd) = mgd$.

Thus $$mgd = \frac{3mgx}{2} - \frac{mg(d + x)}{2} ,$$

$$3mgd = 2mgx \quad \Rightarrow \quad x = \frac{3d}{2} .$$

Therefore the toboggan travels $\dfrac{5d}{2}$ metres down the slope.

10.3 Exercises

10.1 (multiple choice) A particle of mass 2 kg moving in a straight line experiences a constant force of 5 N in the direction of its motion. At a given instant the particle has a speed of 2 m s^{-1}. 4 seconds later the kinetic energy in joules, J, of the particle is

A, 100;   B, 80;   C, 144;   D, 484;   E, 288.

10.2 A uniform rod AB, of length $2l$ and weight W, has its end A in contact with a rough vertical wall and its end B connected by a string BC to a point C vertically above A with $AC = 2l$. The rod is in equilibrium in a vertical plane perpendicular to the wall with the angle $ACB = \alpha$, where $\alpha < 45$. Show that the tension in the string is $W \cos \alpha$ and that the frictional force on the rod at A must act upwards. Show also that the least possible value of the coefficient of friction at A is $\tan \alpha$. Given that $\cos \alpha = 3/4$ and that the string is an elastic one with modulus of elasticity $3W/2$, find the natural length of the string. (L)

10.3 A particle of mass m is attached to one end of an elastic string of natural length a, the other end of which is attached to a fixed point O. When the particle hangs freely the string extends a further distance $a/4$. The particle is placed at O and allowed to fall freely. Find the maximum extension of the string in the subsequent motion. (L)

10.4 Prove that the elastic energy of a light spring of natural length a and modulus of elasticity λ, stretched by an amount x, is $\lambda x^2 /(2a)$.

A trolley of mass m runs down a smooth track of constant inclination $\pi/6$ to the horizontal, carrying at its front a light spring of natural length a and modulus mga/c, where c is constant. When the spring is fully compressed it is of length $a/4$, and it obeys Hooke's law up to this point. After the trolley has travelled a distance b from rest the spring meets a fixed stop. Show that when the spring has been compressed a distance x, where $x < 3a/4$, the speed v of the trolley is given by

$$cv^2 /g = c (b + x) - x^2.$$

Given that $c = a/10$ and $b = 2a$, find the total distance covered by the trolley before it momentarily comes to rest for the first time. (L)

10.5 A light elastic string AB of natural length a and modulus kmg has a particle of mass m attached to B and A is attached to a fixed point. When the particle is hanging freely it is pulled down a distance $a/2$ to C and released from rest. Show that, if $k \leqslant 2$, the particle performs complete simple harmonic oscillations. For the case $k = 4$ find the total time for the particle to reach C again. (SUJB)

10.6 The depth of water in the harbour at Littlebay is assumed to rise and fall with time in simple harmonic motion. At high water it is 15.6 m, at low water 9.2 m. On Saturday, low water is at 14.05 and high water at 20.20. The captain of a freighter is anxious to leave harbour as early on Saturday afternoon as he can. His ship needs at least 14.3 m of water before he can leave. Determine

(a) the earliest time he can leave harbour,

(b) the latest time on Saturday when he can leave.

   If he fails to leave on Saturday he must wait until Monday. What would be the earliest leaving time on Monday?

10.7 A particle of mass m is made to move with simple harmonic motion by the action of a variable force F. If the maximum value of the force is $8\pi^2 m/49$ and the amplitude of movement is 4 m find

(a) the period T of the oscillation,

(b) the speed of the particle at a time $T/8$ after passing through the centre of oscillation.

10.8 Two particles A and B, each of mass 6 kg, move with simple harmonic motion, centre O, with the same period and amplitude in the same straight line. Initially A is at O moving with speed 13 m s^{-1} towards B which is 2 m from O and moving away from O with speed 5 m s^{-1}.

(a) Find the period and amplitude of the motions each correct to three significant figures.

(b) Find the distance of A from O when B is first at rest momentarily.

(c) Calculate the work done by the forces producing the motion of B in moving it from its initial position to a position distant 1 m from O. (SUJB)

10.9 A mass m rests on a horizontal plate which begins to move vertically at time $t = 0$, performing vertical oscillations such that its height x above a datum level is given by $x = a \sin \omega t$, where a and ω are positive constants. Find the force which the mass exerts on the plate at time t. Deduce that the mass will leave the plate if $a\omega^2 > g$. If $a\omega^2 = 2g$ find, in terms of ω, the time at which the mass first leaves the plate.

10.10 A small cubical block of mass 8 m is attached to one end A of a light elastic spring AB of natural length $3a$ and modulus of elasticity $6mg$. The spring and block are at rest on a smooth horizontal table with AB equal to $3a$ and lying perpendicularly to the face to which A is attached. A second block of equal physical dimensions, but of mass m, moving with a speed $(2ga)^{1/2}$ in the direction parallel to BA impinges on the free end B of the spring. $\frac{1}{2}m\left(\sqrt{2ga}\right)^2$

   Assuming that the heavier block is held fixed and that AB remains straight and horizontal in the subsequent motion, determine

(a) the maximum compression of the spring,

(b) the time that elapses between impact and the lighter block first coming to instantaneous rest.

Assuming now that the heavier block is also free to move determine, at the instant when the blocks are first moving instantaneously with the same velocity, the values of

(c) the common velocity of the blocks,

(d) the compression in the spring. (AEB 1982)

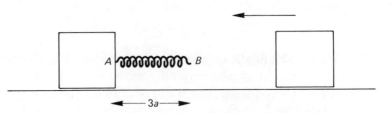

10.11 A particle is projected with initial speed u up a rough slope, inclination α, where $\tan \alpha = \frac{5}{12}$. The coefficient of friction is $\frac{1}{6}$. By considering work done against external forces, find

(a) the distance moved up the slope before coming instantaneously to rest,

(b) the speed of the particle when it returns to its initial position.

10.4 Brief Solutions to Exercises

10.1 Acceleration $= 2.5$ m s^{-2}, $\Rightarrow$ after 4 seconds, velocity $= 12$ m s^{-1} $\Rightarrow$ k.e. $= 144$ J.

<u>Answer C</u>

10.2 $\angle ACB = \angle ABC = \alpha$.

Taking moments about A, $T(2l \sin \alpha) = W(l \sin 2\alpha)$. Thus $T = W \cos \alpha$.

Resolving vertically, $F + T \cos \alpha = W$ $\Rightarrow$ $F = W \sin^2 \alpha$, which is positive $\Rightarrow$ F acts upwards.

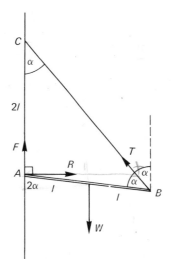

Resolving horizontally, $R = T \sin \alpha = W \sin \alpha \cos \alpha$ $\Rightarrow$ $\dfrac{F}{R} = \tan \alpha$.

Hence $\mu \geqslant \tan \alpha$.

$\cos \alpha = \frac{3}{4}$, $CB = 4l \cos \alpha = 3l$.

Let the natural length of CB be a, then extension $= 3l - a$.

$T = \dfrac{3W}{4} = \dfrac{3W}{2a}(3l - a)$, $a = 6l - 2a$ $\Rightarrow$ $a = 2l$.

[handwritten in right margin:]
$CB^2 = (2l)^2 + (2l)^2 - 2(2l)(2l) \cos(180 - 2\alpha) = CB^2 = 8l^2 + 8l^2 \cos 2\alpha$.
$= 8l^2 + 8l^2 (2\cos^2\alpha - 1)$
$= 8l^2 + 8l^2 \left(\frac{2 \cdot 9}{16} - 1\right)$
$= 8l^2 + 8l^2 \left(\frac{1}{8}\right)$
$= l^2 (8 + 1)$
$= 9l^2$
$CB = 3l$

10.3 $mg = \dfrac{\lambda(a/4)}{a}$ $\Rightarrow$ $\lambda = 4mg$.

At O, e.p.e. (string) $= 0$, k.e. $= 0$

When string has extension x,

$$\text{k.e.} = \tfrac{1}{2}mv^2, \quad \text{e.p.e. (string)} = \frac{2mgx^2}{a}, \quad \text{g.p.e. (particle)} = -mg(a+x).$$

Conservation of energy: $\tfrac{1}{2}mv^2 + \dfrac{2mgx^2}{a} - mg(a+x) = 0$.

When $v = 0$ (lowest point), this gives $x = a$ or $-a/2$.

Since $x > 0$, $x = a$.

Maximum extension in subsequent motion is a.

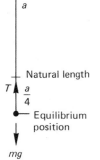

[left margin diagram labels:]
o
$\lambda = \frac{3W}{2}$
a
Natural length
T $\frac{a}{4}$
Equilibrium position
mg

10.4   Bookwork: see Fact Sheet, section 10.1(b).

When compression is x,

Conservation of energy: $\frac{1}{2}mv^2 + \dfrac{mgx^2}{2c} - \dfrac{mg(b+x)}{2} = 0$

$$\Rightarrow \quad cv^2 = gc(b+x) - gx^2.$$

Substituting $c = a/10$, $b = 2a$, $v = 0$ gives: $10x^2 - ax - 2a^2 = 0$

$$\Rightarrow \quad x = -\tfrac{2}{5}a \text{ or } \tfrac{1}{2}a.$$

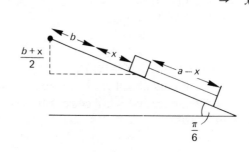

Since $x > 0$, $x = \dfrac{a}{2}$; total distance travelled by trolley $= \dfrac{5a}{2}$.

10.5   Extension in equilibrium $= \dfrac{a}{k} = e$.

When particle is a distance x below equilibrium,

$-m\ddot{x} = T - mg = \dfrac{kmgx}{a}$, which is s.h.m.;

$$x = A\cos\sqrt{\left(\frac{kg}{a}\right)}t = \frac{a}{2}\cos\omega t \text{ where } \omega = \sqrt{\left(\frac{kg}{a}\right)}.$$

Complete s.h.m. oscillations if $\dfrac{a}{2} \leqslant \dfrac{a}{k}$, i.e. $k \leqslant 2$.

If $k = 4$, $e = \dfrac{a}{4}$, $\omega = 2\sqrt{\left(\dfrac{g}{a}\right)}$.

s.h.m. until $x = -\dfrac{a}{4}$,   i.e. $\cos\omega t_1 = -\dfrac{1}{2}$, $t_1 = \dfrac{2\pi}{3\omega} = \dfrac{\pi}{3}\sqrt{\left(\dfrac{a}{g}\right)}$.

When $t = \dfrac{2\pi}{3\omega}$, $|\dot{x}| = \dfrac{a}{2}\omega\sin\omega t_1 = \dfrac{a}{2}\omega\dfrac{\sqrt{3}}{2}$.

Further time to come to instantaneous rest $t_2 = \dfrac{a\omega\sqrt{3}}{4g} = \dfrac{1}{2}\sqrt{\left(\dfrac{3a}{g}\right)}$.

Total time T to return to C is $2(t_1 + t_2)$;

$$T = 2\left(\frac{\pi}{3} + \frac{\sqrt{3}}{2}\right)\sqrt{\left(\frac{a}{g}\right)}$$

$$= \frac{1}{3}\sqrt{\left(\frac{a}{g}\right)}(2\pi + 3\sqrt{3}).$$

10.6 Water depth $x = 12.4 + 3.2 \sin \omega t$ where $\omega = \dfrac{\pi}{375}$;

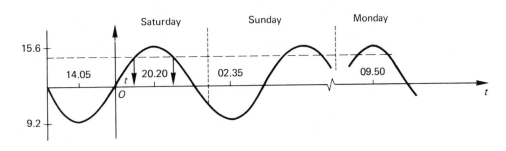

$$x = 14.3 \text{ when } \sin \omega t = \frac{1.9}{3.2}, \ t = 75.9 \text{ minutes after mean water,}$$

i.e. 111 min before and after high tide.

Enough water to leave between 20.20 ± 1 h 51 min, i.e. from 18.29 (a) to 22.11 (b).
Monday morning high tide is at 09.50 so enough water at 07.59.

10.7 s.h.m. has equation of motion $\ddot{x} = -\omega^2 x$ and a solution $x = a \sin \omega t$ where $a = 4$.

Maximum force gives maximum $\ddot{x}$, and so occurs at maximum displacement,

$$\Rightarrow \quad 4\omega^2 = \frac{8\pi^2}{49} .$$

(a) Period $= T = \dfrac{2\pi}{\omega} = 7\sqrt{2}$.

(b) When $t = \dfrac{T}{8} = \dfrac{7\sqrt{2}}{8}$, $\dot{x} = a\omega \cos \omega t = \dfrac{4\pi}{7}$ m s^{-1}.

10.8 A and B have same period and amplitude.

(a) For A, at $t = 0$, $x = 0$, $\dot{x} = 13$; when $x = 2$, $\dot{x} = 5$.

Substituting in $x_A = a \sin \omega t$ gives $\omega = 6$, $a = \dfrac{13}{6}$, so period $= 1.05$ s,

amplitude 2.17 m.

(b) For B, $x_B = \dfrac{13}{6} \sin (6t + \alpha)$, at $t = 0, x = 2$, so $\sin \alpha = \dfrac{12}{13}$.

$\dot{x}_B = 0$ when $6t = \dfrac{\pi}{2} - \alpha$.

At this instant $x_A = \dfrac{13}{6} \sin \left(\dfrac{\pi}{2} - \alpha \right) = \dfrac{5}{6}$.

(c) Work done by forces = change in kinetic energy.
When $x = 1$, $(\dot{x}_B)^2 = \omega^2 (a^2 - x^2) = 133$.
When $x = 2$, $(\dot{x}_B)^2 = 25$.
Work done $= \frac{1}{2} (6) (133 - 25) = 324$ J.

10.9 If reaction of plate on particle is R,

Downward accelerating force $= m\omega^2 x = mg - R \implies R = m(g - \omega^2 x)$.

Leaves plate if $\omega^2 x > g$ for any x, $(-a < x < a)$, i.e. if $a\omega^2 > g$.

If $a\omega^2 = 2g$, particle leaves when $x = \dfrac{a}{2}$, $\implies \sin\omega t = \dfrac{1}{2}$, $\implies t = \dfrac{\pi}{6\omega}$.

10.10 Thrust in spring with compression x is $\dfrac{6mg}{3a}x = \dfrac{2mg}{a}x$; i.e. s.h.m.

(a) Light block moving, retardation $= \dfrac{2g}{a}x$.

$$v\,\dfrac{dv}{dx} = -\dfrac{2gx}{a} \implies \dfrac{v^2}{2} - \dfrac{2ga}{2} = -\dfrac{gx^2}{a};$$

when $v = 0$, $x = a$ (maximum compression).

(b) Periodic time $= 2\pi\sqrt{\left(\dfrac{a}{2g}\right)}$. Time taken to come to rest $= \dfrac{\pi}{2}\sqrt{\left(\dfrac{a}{2g}\right)}$.

(c) Both moving, with same velocity, compression x.

Conservation of energy, $\quad mga = \dfrac{1}{2}(8m + m)v^2 + \dfrac{6mgx^2}{6a}$.

Conservation of momentum, $\quad m\sqrt{(2ga)} = 9mv$.

Solving, $v = \dfrac{\sqrt{(2ga)}}{9}$.

(d) $x = \dfrac{2\sqrt{2}a}{3}$.

10.11 (a) Let total distance moved up the slope $= x$.

$$R = mg\cos\alpha = \dfrac{12}{13}mg, \quad F = \mu R = \dfrac{2mg}{13}.$$

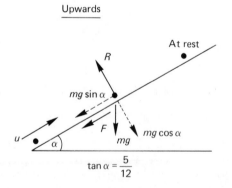

Work done against weight and friction $= \left(mg\sin\alpha + \dfrac{2mg}{13}\right)x$

$$= \dfrac{7mg}{13}x.$$

Therefore $\dfrac{1}{2}mu^2 = \dfrac{7mg}{13}x$, $\quad x = \dfrac{13u^2}{14g}$.

(b)

Downwards

Down slope: gain in k.e. $= \dfrac{5mgx}{13} - \dfrac{2mgx}{13}$,

$$\dfrac{1}{2}mu_1^2 = \dfrac{3mgx}{13} = \dfrac{3mu^2}{14}.$$

Speed of particle at initial position $u_1 = u\sqrt{\left(\dfrac{3}{7}\right)}$.

Thrust in spring with compression $\dfrac{a}{2}$ is $T = \dfrac{\lambda mg}{a}\dfrac{1}{2}\alpha = mg$

T

P.E. stored in spring: $\dfrac{\lambda x^2}{\ell} = \dfrac{\lambda mg}{a}\dfrac{a^2}{2} = \dfrac{mga}{2}$

At max $K.E. = \dfrac{1}{2}mv^2 = mg$

When $T = \dfrac{3}{8}.R$, $V = 0$

$= \dfrac{3}{8}mg$

Put $s =$ distance moved by P from $\dfrac{a}{2}$

$\int s$

153

11 Impulses. Connected Particles

Impulse of a force and the change of momentum. Conservation of momentum. Newton's law of restitution. Direct impact between two particles or spheres. Impact of a particle moving perpendicularly to a plane.

11.1 Fact Sheet

(a) Definitions

    (i) *Momentum* is defined as
mass × velocity = $m\mathbf{v}$. It is a vector.

    (ii) *Impulse* is defined as change in momentum.

$$\mathbf{I} = m\mathbf{v} - m\mathbf{u}, \tag{1}$$

where $\mathbf{u}$ is the velocity before impulse $\mathbf{I}$ and $\mathbf{v}$ is the velocity afterwards.

    (iii) A force $\mathbf{F}$ acting on particle mass $\mathbf{m}$ gives it an *acceleration* $\mathbf{a} = \dfrac{\mathbf{F}}{m}$.

$$\mathbf{v} = \mathbf{u} + \mathbf{a}t \quad \text{so} \quad \mathbf{v} = \mathbf{u} + \frac{\mathbf{F}}{m}t$$

or
$$\mathbf{F}t = m\mathbf{v} - m\mathbf{u}. \tag{2}$$

Comparison of (1) and (2) gives $\mathbf{I} = \mathbf{F}t$, which is an alternative definition of impulse; usually $\mathbf{F}$ is very large and t, the time during which $\mathbf{F}$ acts, is very small.

(b) Conservation of Momentum

When two particles collide, equal but opposite forces act between them for a short time, giving equal but opposite impulses $\mathbf{I}$ and $-\mathbf{I}$.
Assuming motion to be unrestricted:

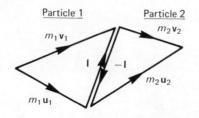

154

For first particle, $\qquad m_1\mathbf{v}_1 - m_1\mathbf{u}_1 = \mathbf{I}$,

For second particle, $\qquad m_2\mathbf{v}_2 - m_2\mathbf{u}_2 = -\mathbf{I}$.

Adding, $\qquad m_1\mathbf{v}_1 + m_2\mathbf{v}_2 = m_1\mathbf{u}_1 + m_2\mathbf{u}_2$.

This is the conservation of momentum equation.

(c) Newton's Law of Restitution

When two particles (1) and (2) are in direct impact then

$$\frac{\text{speed of separation}}{\text{speed of approach}} = e.$$

e is the coefficient of restitution. $0 \leqslant e \leqslant 1$.

If $e = 1$ then the collision is perfectly elastic.

If $e = 0$ then the collision is inelastic and the particles coalesce.

Before: $\quad u_1 \qquad\qquad u_2 \qquad\qquad (u_1 > u_2)$
$\qquad\qquad \rightarrow \qquad\qquad \rightarrow$
$\qquad\qquad (m_1) \qquad\quad (m_2)$

After: $\quad v_1 \qquad\qquad v_2 \qquad\qquad (v_1 < v_2)$
$\qquad\qquad \rightarrow \qquad\qquad \rightarrow$

$$v_2 - v_1 = e\,(u_1 - u_2)$$

(d) Impact of a Particle Moving Perpendicular to a Fixed Plane

Since the plane is fixed the principle of conservation of momentum cannot be applied, but the law of restitution still applies.

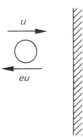

(e) Connected Particles

When particles are connected by a taut string the tension is the same throughout its length even if the string passes over a (frictionless) pulley. When the system is moving, the motion of each particle satisfies $\mathbf{F} = m\mathbf{a}$, the particles having accelerations of equal magnitude.

When a system experiences an impulsive change of state (for example acquiring an extra mass which has a different velocity or if the string suddenly becomes taut), then momentum is conserved through the impulsive tension.

11.2 Worked Examples

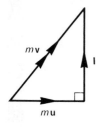

11.1 A particle of mass m is moving with speed u when it receives an impulse of magnitude I at right angles to u. Find the speed and direction of motion of the particle immediately after the impulse and prove that the increase in the kinetic energy of the particle is $\dfrac{I^2}{2m}$.

● Let the velocity of the particle after the impulse **I** be **v**.
   Momentum equation, $m\mathbf{u} + \mathbf{I} = m\mathbf{v}$.
   From the vector triangle, $(mv)^2 = (mu)^2 + I^2$, $\Rightarrow$ $v = \dfrac{\sqrt{(m^2 u^2 + I^2)}}{m}$.

   Direction of motion is at angle $\arctan \dfrac{I}{mu}$ to the direction of u;

   kinetic energy before impulse $= \dfrac{1}{2} mu^2$,

   kinetic energy after impulse $= \dfrac{1}{2} mv^2 = \dfrac{m^2 u^2 + I^2}{2m}$,

   increase in kinetic energy $= \dfrac{1}{2} mu^2 + \dfrac{I^2}{2m} - \dfrac{1}{2} mu^2 = \dfrac{I^2}{2m}$.

11.2 Two small spheres P and Q of equal radius and masses $3m$ and m are placed on a smooth horizontal plane such that PQ is perpendicular to a vertical wall. Q lies between P and the wall. The coefficients of restitution between P and Q and between Q and the wall are both e. If P is projected towards Q (at rest) with a velocity u, find
(a) the speeds of P and Q after they collide;
(b) the speed of Q after rebounding from the wall;
(c) the value of e if P and Q are now travelling in opposite directions at the same speed.

● *P, Q collision:*

$$
\begin{array}{ccc}
 & u & o \\
\text{Before} & \longrightarrow & \longrightarrow \\
 & P\;(3m) & Q\;(m) \\
\text{After} & \longrightarrow & \longrightarrow \\
 & v_P & v_Q
\end{array}
$$

(a) After the collision let P and Q have speeds v_P and v_Q in the direction PQ.

   Conservation of linear momentum: $3mu = 3mv_P + mv_Q$
   $$\Rightarrow \quad 3u = 3v_P + v_Q. \tag{1}$$

   Law of restitution: $eu = v_Q - v_P$ (2)

   $(1) + 3(2)$: $3u + 3eu = 4v_Q$ $\Rightarrow$ $v_Q = \dfrac{3u\,(1+e)}{4}$.

   In (2), $v_P = \dfrac{3u\,(1+e)}{4} - eu = \dfrac{u}{4}\,(3-e)$.

Q, wall collision:
(b) Speed of Q after rebounding from the wall $= ev_Q = \dfrac{3eu(1+e)}{4}$.

156

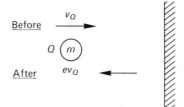

(c) If speed of P = speed of Q,

$$\frac{u}{4}(3 - e) = \frac{3}{4}eu(1 + e)$$

$$\Rightarrow \quad 3 - e = 3e + 3e^2,$$

$$\Rightarrow \quad 3e^2 + 4e - 3 = 0,$$

$$e = \frac{-2 + \sqrt{13}}{3} \quad \left(\text{or } \frac{-2 - \sqrt{13}}{3}, \text{which is impossible}\right).$$

11.3 Three smooth spheres A, B, C, of equal radii and masses m, λm, $\lambda^2 m$ respectively, where λ is a constant, are free to move along a straight horizontal groove with B between A and C. When any two spheres collide the impact is direct and the coefficient of restitution is e. Spheres B and C are initially at rest and sphere A is projected towards sphere B with speed u. Show that the velocities of A and B after the first impact are

$$\frac{1 - \lambda e}{1 + \lambda}u \text{ and } \frac{1 + e}{1 + \lambda}u \text{ respectively.}$$

Find the velocities of B and C after the second impact. Given that $\lambda e < 1$ show that there is a third impact if $e < \lambda$. \hfill (L)

- *A, B collision*

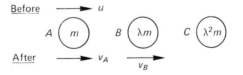

Let the velocities of A and B after the first collision be v_A and v_B.
Conservation of momentum: $\quad mu = mv_A + \lambda mv_B$

$$\Rightarrow \quad u = v_A + \lambda v_B. \tag{1}$$

Law of restitution: $\qquad\qquad eu = v_B - v_A, \tag{2}$

(1) + (2): $\qquad u(1 + e) = v_B(1 + \lambda), \qquad v_B = \frac{u(1 + e)}{(1 + \lambda)}.$

Substitute in (1): $\quad v_A = u - \lambda v_B = \frac{u(1 + \lambda) - \lambda u(1 + e)}{(1 + \lambda)},$

$$v_A = \frac{u(1 - \lambda e)}{(1 + \lambda)}.$$

B, C collision
Let velocities of B and C after collision be u_B and u_C.

Before $\xrightarrow{\ v_B\ }$

$B\left(\lambda m\right)$ $\qquad$ $C\left(\lambda^2 m\right)$

After $\xrightarrow{\ \ }$ $\qquad$ $\xrightarrow{\ \ }$
$\quad u_B$ $\qquad\qquad u_C$

Cons. of momentum: $\quad \lambda mu \dfrac{(1 + e)}{(1 + \lambda)} = \lambda mu_B + \lambda^2 mu_C$

$$\Rightarrow \quad u \frac{(1 + e)}{(1 + \lambda)} = u_B + \lambda u_C. \tag{3}$$

Law of rest.: $\qquad\qquad eu \dfrac{(1 + e)}{(1 + \lambda)} = u_C - u_B. \tag{4}$

These are similar to (1) and (2): solving,

$$u_C = u \frac{(1 + e)^2}{(1 + \lambda)^2}, \qquad u_B = u \frac{(1 + e)(1 - \lambda e)}{(1 + \lambda)^2}.$$

Since $\lambda e < 1$, $\quad v_A$ and u_B are positive and

$$v_A = u_B \left(\frac{1 + \lambda}{1 + e}\right).$$

If $e < \lambda$, $\quad v_A > u_B$, so A and B will collide again.

11.4   A gun of mass 600 kg is free to move along a horizontal track and is connected by a light inelastic rope to an open truck containing sand whose total mass is 1490 kg. The truck is free to move along the same track as the gun. A shell of mass 10 kg is fired from the gun towards the truck and when it leaves the barrel has a horizontal velocity of 915 m s^{-1} relative to the gun and parallel to the track. The shell lodges in the sand where it comes to relative rest before the rope tightens.
   Find,
(a) the speeds of the gun and shell just after the shell leaves the barrel;
(b) the speed of the truck before the rope tightens when the shell is at relative rest inside the truck;
(c) the speed of the gun and truck just after the rope tightens;
(d) the loss in kinetic energy due to the rope tightening;
(e) the magnitude of the impulsive tension in the rope.   (SUJB)

● (a) After the shell leaves the gun let the speeds be v_S and v_G (in opposite directions).

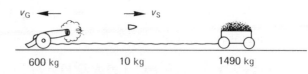

600 kg $\qquad$ 10 kg $\qquad$ 1490 kg

By conservation of linear momentum:

$$10v_S = 600v_G \quad \Rightarrow \quad v_S = 60v_G.$$

Velocity of shell relative to gun $= v_S + v_G = 915$ m s^{-1}.

Solving, $\quad v_G = \dfrac{915}{61} = 15$ m s^{-1}, $\quad v_S = 915 - 15 = 900$ m s^{-1}.

The speeds of the gun and shell are 15 m s^{-1} and 900 m s^{-1} respectively.

(b) When the shell is at rest relative to the truck, let their common speed be v_T.
By conservation of momentum:

$$10\,(900) = 1500 v_T \qquad \text{so} \qquad v_T = 6 \text{ m s}^{-1}.$$

Speed of truck before the rope tightens is 6 m s^{-1}.

(c) Before rope tightens, total momentum in the direction of motion of the gun $= 600\,(15) - 1500\,(6) = 0$ kg m s^{-1}.
Therefore momentum after rope tightens $= 0$ kg m s^{-1}.
Gun and truck are at rest after rope has tightened.

(d) Kinetic energy before rope tightens is

$$\tfrac{1}{2}\,(1500)\,(6)\,(6) + \tfrac{1}{2}\,(600)\,(15)\,(15) = 94.5 \text{ kJ}.$$

Kinetic energy after $= 0$.
Loss in kinetic energy $= 94.5$ kJ

(e) Impulsive tension in the rope $=$ change in momentum of gun (or truck)
$= 9000$ N s.

11.5 A smooth hemispherical bowl of radius a is fixed with its rim horizontal and uppermost. Particle P of mass $3m$ is held at a point A on the rim of the bowl and released from rest so that it moves down inside the bowl. A short time later a second particle Q of mass m, held at B, the other end of the diameter from A, is subjected to a downward vertical impulse of $m\sqrt{(6ga)}$ so that it also moves down the inside of the bowl, reaching the lowest point at the same instant as P.
Find,

(a) the speed of each particle just before collision;

(b) the coefficient of restitution e between P and Q if P is brought to rest by the collision.

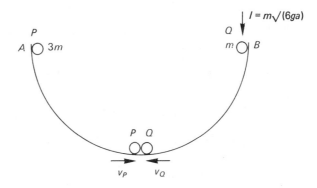

- *Particle P:*

As P descends, loss in potential energy $=$ gain in kinetic energy.

At lowest point,

$$\text{p.e. lost} = 3mga = \tfrac{1}{2}\,(3m)\,(v_P)^2 \text{ giving } v_P = \sqrt{(2ga)}.$$

Particle Q:
Initially, $mu_Q = \text{impulse} = m\sqrt{(6ga)}$ so $u_Q = \sqrt{(6ga)}$.
$\therefore$ initial k.e. is $\tfrac{1}{2}m\,(6ga) = 3mga$.
As Q descends, potential energy lost $= mga$.
At lowest point, k.e. $= 3mga + mga = 4mga = \tfrac{1}{2}mv_Q^2$, giving $v_Q = 2\sqrt{(2ga)}$.

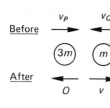

Before v_P v_Q

$3m$ m

After

O v

(a) Just before collision P and Q have speeds $\sqrt{(2ga)}$ and $2\sqrt{(2ga)}$ respectively.
Let Q have speed v after collision.
By conservation of momentum,

$$3m\sqrt{(2ga)} - m2\sqrt{(2ga)} = mv \qquad \Rightarrow \qquad v = \sqrt{(2ga)}.$$

By law of restitution, $e(\sqrt{(2ga)} + 2\sqrt{(2ga)}) = \sqrt{(2ga)}$. Thus $e = \frac{1}{3}$.
(b) Coefficient of restitution $= \frac{1}{3}$.

11.6   A model aircraft takes off at time $t = 0$ s from a point O and climbs at $30\sqrt{2}$ m s^{-1} at $60°$ to the horizontal in a north-easterly direction. Given unit vectors such that **i** points east, **j** points north and **k** points vertically upwards, show that the position vector of the aircraft, relative to O, at time t s is given by $15t\,(\mathbf{i} + \mathbf{j} + \sqrt{6}\mathbf{k})$ m. At time $t = 0$ s, a second model aircraft which is at a height h m above a point 150 m due east of O is moving with constant speed v m s^{-1} in the direction of the vector $\mathbf{j} - \sqrt{3}\mathbf{k}$.
Given that the two aircraft collide, find,
(a) the value of t when the collision takes place,
(b) the value of h and v,
(c) the angle between the directions of motion of the aircraft.
The entangled wreckage is seen to move horizontally immediately after the collision. Find the ratio of the masses of the aircraft.   (AEB 1984)

- *Aircraft 1:* Speed $v_1 = 30\sqrt{2}$ at $60°$ to horizontal.
Vertical component $= 30\sqrt{2}\cos 30 = 15\sqrt{6}$.
Horizontal component $= 30\sqrt{2}\sin 30 = 15\sqrt{2}$.
This is in a north-easterly direction so has components $15\sqrt{2}\cos 45 = 15$ due east and due north.
Therefore $\mathbf{v}_1 = 15(\mathbf{i} + \mathbf{j} + \sqrt{6}\mathbf{k})$ and position vector from
$O = \mathbf{r}_1 = 15t\,(\mathbf{i} + \mathbf{j} + \sqrt{6}\mathbf{k})$.

Aircraft 2: Speed is in direction $\mathbf{j} - \sqrt{3}\mathbf{k}$.
A unit vector in this direction is $\frac{1}{2}\mathbf{j} - \frac{\sqrt{3}}{2}\mathbf{k}$.

$$\mathbf{v}_2 = v\left(\frac{1}{2}\mathbf{j} - \frac{\sqrt{3}}{2}\mathbf{k}\right) \text{ and } \mathbf{r}_2 = \frac{vt}{2}(1\mathbf{j} - \sqrt{3}\mathbf{k}) + \mathbf{C}.$$

When $t = 0$, $\mathbf{r}_2 = 150\mathbf{i} + h\mathbf{k} = \mathbf{C}$, so

$$\mathbf{r}_2 = 150\mathbf{i} + \frac{vt}{2}\mathbf{j} + \left(h - \frac{v\sqrt{3}t}{2}\right)\mathbf{k}.$$

When the planes collide, $\mathbf{r}_1 = \mathbf{r}_2$.
Equate components,
(i) $15t = 150$, $t = 10$;

(j) $15t = \frac{v}{2}t$, $v = 30$;

(k) $15\sqrt{6}t = h - \frac{v\sqrt{3}t}{2}$, $h = 150\sqrt{6} + 150\sqrt{3} = 150(\sqrt{6} + \sqrt{3})$.

(a) Collision takes place after 10 s.

(b) $h = 150(\sqrt{6} + \sqrt{3})$ m, $v = 30$ m s^{-1}.

(c) $|\mathbf{v}_1| = 30\sqrt{2}$, $|\mathbf{v}_2| = v = 30$.
If θ is the angle between the directions of motion, using scalar products,
$\mathbf{v}_1 . \mathbf{v}_2 = |\mathbf{v}_1||\mathbf{v}_2|\cos\theta \Rightarrow 225 - 225\sqrt{(18)} = 30\sqrt{2}(30)\,(\cos\theta)$.
$\cos\theta = -0.5732 \Rightarrow \theta = 125.0°$.

Let the aircraft masses be m_1 and m_2.

Since the wreckage moves horizontally after the collision, the vertical component **(k)** of the momentum must be zero.

This requires $m_1(15\sqrt{6}) + m_2\left(-\dfrac{30\sqrt{3}}{2}\right) = 0$,

i.e. $\dfrac{m_1}{m_2} = \dfrac{30\sqrt{3}}{30\sqrt{6}}$ so $m_1 : m_2 = 1 : \sqrt{2}$.

11.7 $n+1$ beads of equal mass, $A_0, A_1, \ldots, A_n$, are at rest at distances $0, a, \ldots,$ na respectively from a fixed point on a smooth horizontal straight wire. The bead A_0 is projected towards A_1 with speed u_0. For all impacts the coefficient of restitution is e (< 1). Prove that A_1 begins to move with speed $\frac{1}{2}(1+e)u_0$.

Find the speed with which A_n begins to move and the time at which this occurs after the projection of A_0. (OLE)

● Let the speeds of A_1 and A_0 after the first collision be u_1 and v_0, respectively.

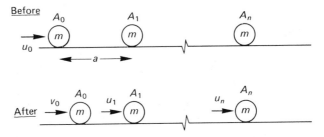

Conservation of momentum: $mu_0 = mv_0 + mu_1 \Rightarrow u_0 = v_0 + u_1$.
Law of restitution: $eu_0 = u_1 - v_0$.

Adding, $u_0(1+e) = 2u_1$ so $u_1 = \dfrac{u_0}{2}(1+e)$.

Similarly, $u_2 = \dfrac{u_1}{2}(1+e) = u_0\left(\dfrac{1+e}{2}\right)^2$, etc.

A_n moves with speed $u_n = u_0\left(\dfrac{1+e}{2}\right)^n$.

Time taken for A_0 to reach A_1 is $\dfrac{a}{u_0}$,

for A_1 to reach A_2 is $\dfrac{a}{u_1} = \dfrac{a}{u_0}\left(\dfrac{2}{1+e}\right)$, etc.

Total time until A_{n-1} reaches A_n

$$= \dfrac{a}{u_0}\left[1 + \dfrac{2}{1+e} + \left(\dfrac{2}{1+e}\right)^2 + \ldots + \left(\dfrac{2}{1+e}\right)^{n-1}\right].$$

This is a geometric progression with $r = \dfrac{2}{1+e}$.

Total time $= \dfrac{a}{u_0}\dfrac{(r^n - 1)}{(r-1)}$ where $r - 1 = \dfrac{2}{1+e} - 1 = \dfrac{1-e}{1+e}$.

Total time $= \dfrac{a}{u_0}\dfrac{(1+e)}{(1-e)}\left[\left(\dfrac{2}{1+e}\right)^n - 1\right]$.

11.8 A particle of mass $2m$ is on a plane inclined at an angle $\tan^{-1} 3/4$ to the horizontal. The particle is attached to one end of a light inextensible string. This string runs parallel to a line of greatest slope of the plane, passes over a small smooth pulley at the top of the plane and then hangs vertically carrying a particle

161

of mass $3m$ at its other end. The system is released from rest with the string taut. Find the acceleration of each particle and the tension in the string when the particles are moving freely, given that

(a) the plane is smooth,

(b) the plane is rough and the coefficient of friction between the particle and the plane is 1/4.   (L)

• $\tan \alpha = \frac{3}{4}$, $\sin \alpha = \frac{3}{5}$, $\cos \alpha = \frac{4}{5}$.

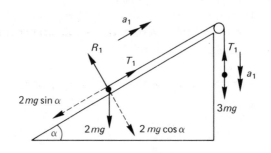

For each particle $F = ma$.

(a) *Smooth plane:*

Let the magnitude of acceleration of each particle be a_1.

For the particle on the plane, $T_1 - 2mg \sin \alpha = 2ma_1$ $\Rightarrow$ $T_1 - \frac{6}{5}mg = 2ma_1$.

For the hanging particle, $3mg - T_1 = 3ma_1$.

Add: $3mg - \frac{6}{5}mg = 5ma_1$ $\Rightarrow$ $a_1 = \frac{9}{25}g$ and $T_1 = 3m(g - a_1) = \frac{48}{25}mg$.

Acceleration of each particle is $\frac{9}{25}g$ and the tension in the string is $\frac{48}{25}mg$.

(b) *Rough plane:*

Let the magnitude of acceleration of each particle be a_2.

For the particle on the plane,

$$R_2 = 2mg \cos \alpha = \frac{8}{5}mg; \qquad F_2 = \mu R_2 \qquad \text{so} \qquad F_2 = \frac{2}{5}mg.$$

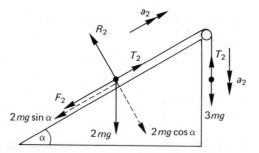

Resolving up the plane, $T_2 - \frac{6}{5}mg - \frac{2}{5}mg = 2ma_2$.

For the hanging particle, $3mg - T_2 = 3ma_2$.

Add: $3mg - \frac{8}{5}mg = 5ma_2$ $\Rightarrow$ $a_2 = \frac{7}{25}g$, $T_2 = 3m(g - a_2) = \frac{54}{25}mg$.

Acceleration of each particle is $\frac{7}{25}g$ and the tension in the string is $\frac{54}{25}mg$.

11.9   Two particles of masses $4m$ and $3m$ respectively are attached one to each end of a light inextensible string which passes over a small smooth pulley. The particles move in a vertical plane with both hanging parts of the string vertical. Write down the equation of motion for each of the particles and hence determine, in terms of m and/or g as appropriate, the magnitude of the acceleration of the particles and of the tension in the string.

When the particle of mass $3m$ is moving upwards with a speed V it picks up from rest at a point A an additional mass $2m$ so as to form a composite particle Q of mass $5m$. Determine

(a) the initial speed of the system,

(b) the impulsive tension in the string immediately the additional particle has been picked up,

(c) the height above A to which Q rises.   (AEB 1983)

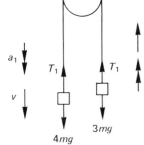

● Let the magnitude of the common acceleration be a_1.
Particle of mass $4m$: $4mg - T_1 = 4ma_1$.
Particle of mass $3m$: $T_1 - 3mg = 3ma_1$.

Add: $mg = 7ma_1$ $\Rightarrow$ $a_1 = \dfrac{g}{7}$.

Substituting, $T_1 = 3m(a_1 + g) = \dfrac{24}{7} mg$.

Particles have an acceleration $\dfrac{g}{7}$ and the string has a tension $\dfrac{24}{7} mg$.

When $3m$ picks up $2m$, the total momentum of the whole system is conserved, so if V_1 is the speed after the pickup,

$$4mV + 3mV = (4m + 5m) V_1 \quad \Rightarrow \quad V_1 = \frac{7}{9} V. \tag{a}$$

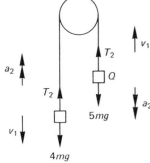

To find impulsive tension I consider the change in momentum of mass $4m$:

$$|4mV_1 - 4mV| = I, \quad \Rightarrow \quad I = 4mV - 4m\left(\frac{7}{9} V\right) = \frac{8}{9} mV. \tag{b}$$

Let composite particle Q rise a distance h from A.
Equations of motion are $T_2 - 4mg = 4ma_2$
and $5mg - T_2 = 5ma_2$.

Solving, $a_2 = \dfrac{g}{9}$ where a_2 is a retardation.

Using $v^2 = u^2 + 2as$, $0 = \dfrac{49}{81} V^2 - \dfrac{2}{9} gh$ $\Rightarrow$ $h = \dfrac{49V^2}{18g}$.

Q rises a distance $\dfrac{49V^2}{18g}$ above A.   (c)

Alternatively for (c): The equation of conservation of energy may be used, remembering that while the particle of mass $5m$ is gaining potential energy $5mgh$, the particle of mass $4m$ is losing $4mgh$.

$$5mgh - 4mgh = \frac{1}{2}(5m) V_1^2 + \frac{1}{2}(4m) V_1^2, \quad \Rightarrow \quad mgh = \frac{9}{2}m V_1^2;$$

$$h = \frac{9V_1^2}{2g} = \frac{9(49V^2)}{2g(81)} = \frac{49V^2}{18g}.$$

11.10 A particle A of mass m is placed on a rough plane inclined at an angle $\tan^{-1}(3/4)$ to the horizontal. One end of a light inextensible string is attached to A and the string runs parallel to a line of greatest slope of the plane, passes over a small smooth pulley at the top of the plane and has a second identical particle B attached to its other end. The particles are released from rest when the string is taut and the part of the string between B and the pulley is vertical. Given that the coefficient of friction between A and the plane is 1/4, show that B moves with acceleration $g/10$ and find the tension in the string.

After time t_0 the string breaks when A is at the point P on the plane and A comes momentarily to rest at a point Q on the plane. Show that $PQ = gt_0^2/160$.

(L)

● Let the magnitude of the acceleration of each particle be a.

For particle A:

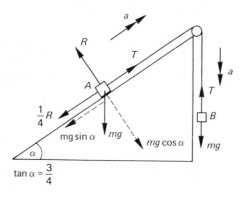

$$\tan \alpha = \frac{3}{4}$$

$R = mg \cos \alpha = \tfrac{4}{5} mg, \quad F = \mu R = \tfrac{1}{5} mg.$

Resolving up the slope, $\quad T - mg \sin \alpha - F = ma,$

$$T - \tfrac{3}{5} mg - \tfrac{1}{5} mg = ma,$$

$$T - \tfrac{4}{5} mg = ma. \tag{1}$$

For particle B: $\quad mg - T = ma.$ (2)

Adding (1) and (2): $\quad \tfrac{1}{5} mg = 2ma \quad \Rightarrow \quad a = \dfrac{g}{10}.$

From (2), $\qquad\qquad\qquad T = m(g - a) = \dfrac{9mg}{10}.$

The particles have an acceleration of $\dfrac{g}{10}$ and the tension in the string is $\dfrac{9mg}{10}.$

At time t_0, particle A, at P, has velocity $\dfrac{gt_0}{10}$ up the plane.

Then the string breaks, $T = 0$,

(1) becomes $ma = -\dfrac{4mg}{5}$ so A has an acceleration $-\dfrac{4g}{5}.$

If particle moves a distance s up the plane before coming to rest at Q,

$v^2 = u^2 + 2as$ gives

$$\left(\frac{gt_0}{10}\right)^2 = 2\left(\frac{4g}{5}\right)s, \quad \text{so} \quad s = \frac{(5g^2)(t_0^2)}{(8g)(100)} = \frac{gt_0^2}{160}. \quad PQ = \frac{gt_0^2}{160}.$$

11.3 Exercises

11.1   A sphere A of mass $5m$ and moving with speed u collides with a sphere B, of the same radius and of mass $2m$, moving in the opposite direction with speed $2u$. In the impact sphere A is brought to rest. Calculate the coefficient of restitution between the spheres and the loss of kinetic energy in the collision.

11.2   Clear diagrams showing velocities before and after impact must be drawn.
   A and B are two small smooth spheres of equal radii and masses m and $2m$

respectively, which lie at rest on a horizontal plane, their line of centres being perpendicular to a vertical barrier. A is projected towards B with speed u and collides with B; B goes on to strike the barrier from which it rebounds to strike A again. This collision reduces B to rest. If the coefficient of restitution between A and B is 1/4, find:

(a) the coefficient of restitution between B and the barrier;

(b) the final speed of A in terms of u;

(c) the magnitude of the impulse that A exerts on B during the second collision of the spheres;

(d) the total loss of energy due to the three impacts.   (SUJB)

11.3 Two particles P and Q of equal mass m are positioned, as shown in the diagram, on a smooth horizontal surface between two parallel vertical walls, the

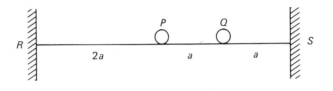

line joining the particles being perpendicular to the walls. The coefficient of restitution between the particles is 1/3 and between the particles and the walls is 1/2.

If P is projected towards R with speed $2u$ and Q is projected towards S with speed u, find;

(a) the time taken until P and Q collide with each other,

(b) the speed and direction of P and Q after they have collided.

11.4 A boy A of mass m is standing at rest on roller skates on a smooth horizontal surface. He is holding a ball B of mass km which he throws to a friend standing directly in front of him. If the energy expended by the boy is E find the velocities of the boy and the ball in terms of m, k and E immediately after the ball is released.

11.5 A small smooth sphere moves on a horizontal table and strikes an identical sphere lying at rest on the table at a distance d from a vertical wall, the impact being along the line of centres and perpendicular to the wall. Prove that the next impact between the two spheres will take place at a distance

$$2de^2/(1 + e^2)$$

from the wall, where e is the coefficient of restitution for all impacts involved.
          (L)

11.6 Clear diagrams showing velocities before and after impact must be drawn.

Two small spheres, A and B, have masses $3m$ and m respectively. They are placed on a smooth floor so that the line joining their centres is perpendicular to a wall. The sphere B is a distance x from the wall, and is the nearer of the two spheres to the wall. The coefficient of restitution between A and B is e, and between B and the wall is $\frac{1}{3}$. Sphere A is projected with velocity u and hits B directly; B then hits the wall, rebounds and hits A again.

Prove that

(a) when B hits the wall, A is a distance $\dfrac{4ex}{3(1+e)}$ from the wall;

(b) when B rebounds from the wall, A and B approach one another with a speed of u;

(c) the time interval between the first impact of A on B and the second impact of A on B is $\dfrac{4x}{3u}$.

What is the surprising feature of the result in part (c)? (SUJB)

11.7 Two particles, of masses 1 kg and $\sqrt{3}$ kg, are attached to the ends A and B respectively of a light inextensible string which passes over a smooth pulley C at the top of a smooth plane inclined at an angle of 60° to the horizontal. Initially the particles are at rest with the string taut, the portion AC hanging vertically and BC lying along the line of greatest slope of the plane. When the system is released B moves a distance 2 m on the slope before striking a horizontal plane.

Find (a) the total distance moved by A vertically, before coming instantaneously to rest;

     (b) the impulsive tension in the string when the string again becomes taut.

11.8 A particle P of mass $8m$ rests on a smooth horizontal rectangular table and is attached by light inelastic strings to particles Q and R of mass $2m$ and $6m$ respectively. The strings pass over light smooth pulleys on opposite edges of the table so that Q and R can hang freely with the strings perpendicular to the table edges. The system is released from rest. Obtain the equations of motion of each of the particles and hence, or otherwise, determine the magnitude of their common acceleration and the tensions in the strings.

After falling a distance x from rest, the particle R strikes an inelastic floor and it is brought to rest. Determine the further distance y that Q ascends before momentarily coming to rest. [It is to be assumed that the lengths of the strings are such that P remains on the table and Q does not reach it.] (AEB 1984)

11.9 In the diagram, 3 particles of masses m_1, m_2 and m_3 are attached at points A, B and C of an inextensible string which passes over smooth pulleys at Q and

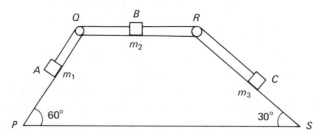

R. Plane PQ is smooth, QR rough with coefficient of friction $\frac{1}{4}$, RS rough with coefficient of friction $1/\sqrt{3}$.

The system begins to move with the particle at A moving down the slope. Find the acceleration of the system. If $m_1 = 4\sqrt{3}$ kg, $m_2 = 4$ kg and $m_3 = 3$ kg, find the tensions in the two parts of the string. (Take g as 10 m s^{-2}.) Find the work done against frictional forces when the particles have each moved 0.5 m.

11.10 Two particles A and B, of masses $2m$ and m, are attached to the ends of a light inextensible string which passes over a fixed light smooth pulley. The system is released from rest with both portions of the string vertical and taut, and A and B on the same level. Find the acceleration of either particle and the tension in the string. The string breaks when the speed of either particle is V. Find the distance between the particles when B reaches its highest point, assuming that A does not strike a surface or B reach the pulley.

11.4 Brief Solutions to Exercises

11.1 Cons. of mtm: $5mu - 4mu = 2mv \Rightarrow v = \dfrac{u}{2}$.

Before → u → $-2u$

A $\bigcirc$ $5m$ B $\bigcirc$ $2m$

After → v

Law of rest.: $e(3u) = v = \dfrac{u}{2} \Rightarrow e = \frac{1}{6}$.

Loss of k.e. $= \dfrac{1}{2}(5m)u^2 + \dfrac{1}{2}(2m)(2u)^2 - \dfrac{1}{2}(2m)\left(\dfrac{u}{2}\right)^2 = \dfrac{25}{4}mu^2$.

11.2 *Collision 1:*

← u Before

$\bigcirc$ $2m$ B $\bigcirc$ m A

← v_2 ← v_1 After

Cons. of mtm: $mu = mv_1 + 2mv_2$.

Law of rest.: $\dfrac{1}{4}u = v_2 - v_1$, so $v_2 = \dfrac{5u}{12}$, $v_1 = \dfrac{u}{6}$.

A and B both move towards barrier.

B rebounds with speed $e\left(\dfrac{5u}{12}\right)$ towards A.

Collision 2:

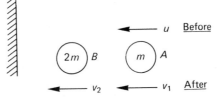

$\dfrac{5ue}{12}$ $\dfrac{u}{6}$

→   →   Before

$\bigcirc$ $2m$ B $\bigcirc$ m A

→   →   After

O v_3

Cons. of mtm: $2m\left(\dfrac{5eu}{12}\right) - m\left(\dfrac{u}{6}\right) = mv_3 \Rightarrow v_3 = \dfrac{u}{6}(5e - 1)$.

Law of rest.: $\dfrac{1}{4}\left(\dfrac{5eu}{12} + \dfrac{u}{6}\right) = v_3 = \dfrac{u}{6}(5e - 1) \Rightarrow e = \dfrac{2}{7}$. (a)

Final speed of A is $v_3 = \dfrac{u}{6}\left[5\left(\dfrac{2}{7}\right) - 1\right] = \dfrac{u}{14}$. (b)

Impulse $= 2m\left(\dfrac{5eu}{12}\right) = \dfrac{5mu}{21}$. (c)

Loss of k.e. $= \dfrac{1}{2}mu^2 - \dfrac{1}{2}m\left(\dfrac{u}{14}\right)^2 = \dfrac{195}{392}mu^2$. (d)

11.3 P, speed $2u$, travels $2a$ in time $\dfrac{a}{u}$.

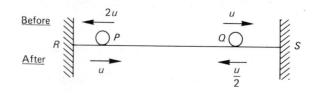

Q, speed u, travels a in time $\dfrac{a}{u}$.

P and Q strike walls at the same time and rebound with speeds u and $\dfrac{u}{2}$ respectively,

closing speed = $\dfrac{3u}{2}$.

(a) Total time to collision = $\dfrac{a}{u} + 4a\left(\dfrac{2}{3u}\right) = \dfrac{11a}{3u}$.

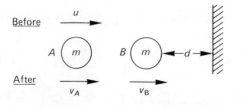

(b) Cons. of mtm: $mu - m\dfrac{u}{2} = mv_1 + mv_2$.

Law of rest.: $\dfrac{1}{3}\left(u + \dfrac{u}{2}\right) = v_2 - v_1$,   hence $v_1 = 0$, $v_2 = \dfrac{u}{2}$.

P stops, Q rebounds with speed $\dfrac{u}{2}$.

11.4   Let velocities after ball is thrown be v_A and v_B towards the friend.
Cons. of mtm: $mv_A + kmv_B = 0$.
Kinetic energy = $E = \frac{1}{2}mv_A^2 + \frac{1}{2}kmv_B^2$,

Solving, $v_A = -\sqrt{\left(\dfrac{2Ek}{m(1+k)}\right)}$, $v_B = \sqrt{\left(\dfrac{2E}{mk(1+k)}\right)}$.

11.5   Cons. of mtm: $mu = mv_A + mv_B$.

Law of rest.: $eu = v_B - v_A$, $\Rightarrow$ $v_A = \dfrac{u}{2}(1-e)$, $v_B = \dfrac{u}{2}(1+e)$.

Time for B to reach wall = $\dfrac{2d}{u(1+e)}$;

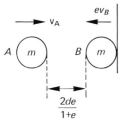

A is then $\dfrac{2de}{(1+e)}$ from B. (1)

After rebound B has speed $e\dfrac{u}{2}(1+e)$ towards A.

Closing speed $= \dfrac{u}{2}(1+e^2)$, (2)

Further time to collision $= (1)/(2)$, B is then $\dfrac{2de^2}{1+e^2}$ from wall.

11.6 Cons. of mtm: $3u = 3v_A + v_B$.

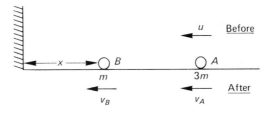

Law of rest.: $eu = v_B - v_A$, $\Rightarrow$ $v_A = \dfrac{u}{4}(3-e)$, $v_B = \dfrac{3u}{4}(1+e)$.

Time for B to reach wall $= \dfrac{4x}{3u\,(1+e)}$.

(a) When B reaches wall, A is distance $\dfrac{4ex}{3\,(1+e)}$ away.

(b) After rebound approach speed $= \dfrac{u}{4}(3-e) + \dfrac{1}{3}\left(\dfrac{3u}{4}\right)(1+e) = u$.

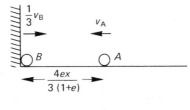

(c) Total time to next collision:

$$\dfrac{4x}{3\,(1+e)\,u} + \dfrac{4ex}{3\,(1+e)\,u} = \dfrac{4x}{3u}\ . \text{ Independent of } e!$$

11.7
For A, $\quad T - g = a$,

for B, $\quad \sqrt{3}g\left(\dfrac{\sqrt{3}}{2}\right) - T = \sqrt{3}a$.

$a = \dfrac{g}{2\,(\sqrt{3}+1)} = 1.83 \text{ m s}^{-2}$.

When B strikes plane, $\quad s = 2$, $\quad v^2 = u^2 + 2as$ gives $\quad v = 2.71 \text{ m s}^{-1}$.

A now moves up under gravity a further distance h where $2gh = v^2 \Rightarrow h = 0.366$ m.
(a) Total distance moved by A = 2.37 m.
 Just before string again becomes taut $v_A = 2.71 \text{ m s}^{-1}$.

Cons. of mtm: $1v + \sqrt{3}v = 2.71$ $\Rightarrow$ $v = 0.99$ m s^{-1}.

(b) Impulsive tension $= \sqrt{3}v = 1.72$ N s.

11.8   Acceleration $= a$.

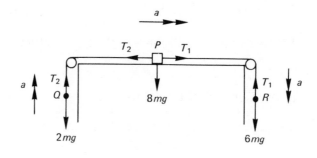

For R, $6mg - T_1 = 6ma$;

for P, $T_1 - T_2 = 8ma$;   for Q, $T_2 - 2mg = 2ma$.

Add: $a = \dfrac{g}{4}$,   so $T_1 = \dfrac{9}{2}mg$, $T_2 = \dfrac{5}{2}mg$.

When R has fallen a distance x, $v^2 = u^2 + 2as$ gives $v = \sqrt{\left(\dfrac{gx}{2}\right)}$

When $T_1 = 0$, acceleration of Q is $\dfrac{-g}{5}$.

Using $v^2 = u^2 + 2as$, $0 = \dfrac{gx}{2} - \dfrac{2gy}{5}$, $y = \dfrac{5x}{4}$,

Q ascends a further distance $\dfrac{5x}{4}$.

11.9

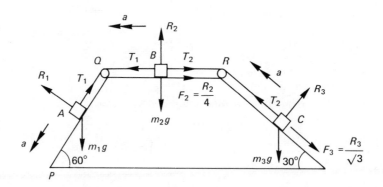

At A, $m_1 g \sin 60 - T_1 = m_1 a$;

at B, $F_2 = \dfrac{m_2 g}{4}$,   so $T_1 - T_2 - \dfrac{m_2 g}{4} = m_2 a$;

at C, $R_3 = m_3 g \cos 30$, $F_3 = \dfrac{m_3 g}{2}$, $T_2 - m_3 g \sin 30 - \dfrac{m_3 g}{2} = m_3 a$.

Adding, $a = \dfrac{g(2\sqrt{3}m_1 - m_2 - 4m_3)}{4(m_1 + m_2 + m_3)}$.

With $m_1 = 4\sqrt{3}$, $m_2 = 4$, $m_3 = 3$, $a = \dfrac{2g}{7 + 4\sqrt{3}}$ m s^{-2} = 1.436 m s^{-2}.

$T_1 = 50.1$ N, $T_2 = 34.3$ N.

Work done against friction $= (F_2 + F_3)(0.5) = \frac{5}{4}g$ J.   (12.5 J.)

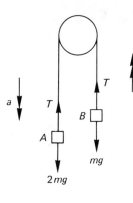

11.10   For A, $2mg - T = 2ma$,

for B, $T - mg = ma$,   so $a = \dfrac{g}{3}$ and $T = \dfrac{4mg}{3}$.

At speed V, using $v^2 = u^2 + 2as$, A and B have each moved a distance $\dfrac{3V^2}{2g}$ when string breaks.

Moving under gravity:

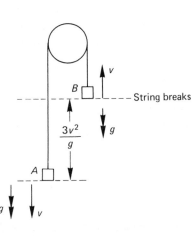

Particle B: Using $v^2 = u^2 - 2gs_1$, $s_1 = \dfrac{V^2}{2g}$,   time taken $= \dfrac{V}{g}$.

Total distance risen by $B = \dfrac{3V^2}{2g} + \dfrac{V^2}{2g} = \dfrac{2V^2}{g}$.

Particle A: Using $s = ut + \dfrac{1}{2}gt^2$, $u = V$, $t = \dfrac{V}{g}$,   so $s_2 = \dfrac{3V^2}{2g}$.

Total distance fallen by $A = \dfrac{3V^2}{g}$,   distance $AB = \dfrac{5V^2}{g}$.

12 Probability

Simple applications of elementary probability.
Conditional probability, mutually exclusive, exhaustive and independent events.
Sum and product laws.
The meaning and use of simple tree diagrams.

12.1 Fact Sheet

(a) For one event A

(i) $0 \leqslant P(A) \leqslant 1$;

(ii) A' or $\overline{A}$ denotes the event 'A does not happen', and $P(A) + P(A') = 1$.

(b) For two events, A and B

(i) $P(A$ or $B)$ means A and/or B occurs and is written $P(A \cup B)$;

(ii) $P(A$ and $B)$ means that both A and B occur and is written $P(A \cap B)$;

(iii) $P(A \cup B) = P(A) + P(B) - P(A \cap B)$.

(c) Conditional Probability

The probability of A, given that B has already occurred, is written $P(A \mid B)$

$$P(A \mid B) = \frac{P(A \cap B)}{P(B)}, P(B) \neq 0.$$

(d) Independent Events

If $P(A \mid B) = P(A)$ then the occurrence or non-occurrence of event B does not influence the probability of event A. The events are said to be independent and $P(A) \cdot P(B) = P(A \cap B)$.

This is the multiplication law for independent events.

(e) Mutually Exclusive Events

If $P(A \cap B) = 0$ then both of A and B cannot occur. The events are said to be mutually exclusive and $P(A \cup B) = P(A) + P(B)$.

This is the addition law for mutually exclusive events.

(f) Exhaustive Events

$P(A \cup B) = 1.$

(g) Tree Diagrams

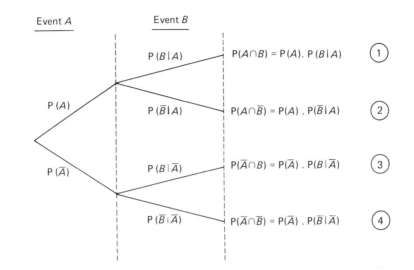

Remember:
 (i) The total probability for any one set of 'branches' = 1.
 (ii) The sum of final probabilities (intersections) = 1.
 (iii) The tree shows conditional probabilities for $(B \mid A)$ etc.
   If $P(A \mid B)$ is required, then:
   (a) Look in the final column for the intersections containing B; in this case $P(B) = ① + ③$.
   (b) Find the term giving $A \cap B$, in this case ①.

   Then $P(A \mid B) = \dfrac{P(A \cap B)}{P(B)} = \dfrac{①}{① + ③}.$

12.2 Worked Examples

12.1   The probability that a fisherman catches a fish is $\dfrac{7}{10}$ on a cloudy day, and $\dfrac{1}{5}$ on a clear day. If the probability of a cloudy day is $\dfrac{3}{5}$, find the probability that the day was cloudy given that he did not catch a fish.

● Let event C be 'the weather was cloudy'.
Let event F be 'he caught a fish'.
To find $P(C \mid \overline{F})$:
P (he did not catch a fish) = $P(\overline{F}) = P(C \cap \overline{F}) + P(\overline{C} \cap \overline{F})$

$\qquad\qquad\qquad = \dfrac{9}{50} + \dfrac{8}{25}$

$\qquad\qquad\qquad = \dfrac{1}{2}.$

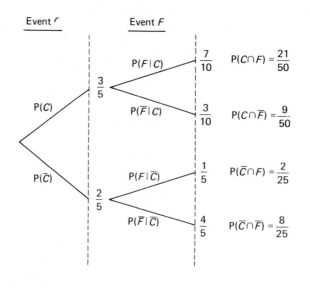

$$P(C \mid \overline{F}) = \frac{P(C \cap \overline{F})}{P(\overline{F})} = \frac{9/50}{1/2} = \frac{9}{25}.$$

12.2 The probability that a man and his wife will fail their driving tests at the first attempt are $\frac{2}{3}$ and $\frac{3}{5}$ respectively. Assuming that these events are independent, determine the probability that

(a) both pass the test at the first attempt,

(b) at least one passes the test at the first attempt.

● Prob. that man fails = 2/3;   prob. that man passes = 1/3;
  prob. that wife fails = 3/5;   prob. that wife passes = 2/5.
  Independent events so,
  (a) prob. that both pass at the first attempt

$$= \left(\frac{1}{3}\right)\left(\frac{2}{5}\right) = \frac{2}{15};$$

(b) prob. that at least one passes

$$= 1 - (\text{prob. that neither passes})$$

$$= 1 - \left(\frac{2}{3}\right)\left(\frac{3}{5}\right) = \frac{3}{5}.$$

12.3 Three boxes, X, Y and Z contain coloured balls. X contains 5 black and 4 white balls, Y contains 7 black and 5 white balls and Z contains 3 black and 5 white balls.

(a) If balls are withdrawn from box Z, with replacement, find the probability that the third ball drawn is the second white ball.

(b) One of the boxes is selected at random and a ball is withdrawn from it. Find the probability that
   (i)   box X was chosen and the ball was black,
   (ii) a white ball was chosen,
   (iii) the ball was selected from box Z, given that it was black.

● (a) Box Z.   P (black drawn) = $\dfrac{3}{8}$,   P (white drawn) = $\dfrac{5}{8}$.

For the third ball to be the second white one the possible orders are BWW and WBW.

The balls are replaced after selection so

174

$$P(BWW) = \left(\frac{3}{8}\right)\left(\frac{5}{8}\right)\left(\frac{5}{8}\right) = \frac{75}{512}, \qquad P(WBW) = \left(\frac{5}{8}\right)\left(\frac{3}{8}\right)\left(\frac{5}{8}\right) = \frac{75}{512}.$$

Prob. that third ball is the second white one $= \dfrac{75}{256}$.

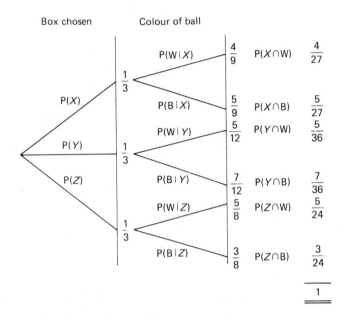

(b) (i) Prob. of box X and a black ball $= P(X \cap B)$

$$= \left(\frac{1}{3}\right)\left(\frac{5}{9}\right) = \frac{5}{27}.$$

(ii) Prob. that a white ball was chosen

$$= P(X \cap W) + P(Y \cap W) + P(Z \cap W)$$

$$= \left(\frac{1}{3}\right)\left(\frac{4}{9}\right) + \left(\frac{1}{3}\right)\left(\frac{5}{12}\right) + \left(\frac{1}{3}\right)\left(\frac{5}{8}\right) = \frac{107}{216}.$$

(iii) Prob. that a black ball was chosen $= 1 - P(W)$

$$= \frac{109}{216}.$$

Prob. of box Z and a black ball $= \left(\frac{1}{3}\right)\left(\frac{3}{8}\right) = \frac{1}{8}$.

Prob. that the ball was selected from box Z given that it was a black ball

$$= P(Z \mid B) = \frac{P(Z \cap B)}{P(B)} = \frac{1/8}{109/216} = \frac{27}{109}.$$

12.4 A card is drawn from a standard pack of 52 playing cards and is not re-placed. A second card is then drawn. Find
(a) the probability that the second card is a club given that the first card is a club.
(b) the probability that the second card is a club.
(c) the expected number of club cards in two draws.
  Answer the same questions again if the first card is replaced before the second card is drawn.

● In a pack of 52 cards, 13 are clubs.

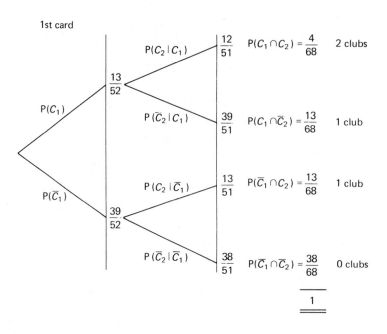

1st card

$P(C_2|C_1)$ $\frac{12}{51}$ $P(C_1 \cap C_2) = \frac{4}{68}$ 2 clubs

$\frac{13}{52}$

$P(C_1)$

$P(\overline{C}_2|C_1)$ $\frac{39}{51}$ $P(C_1 \cap \overline{C}_2) = \frac{13}{68}$ 1 club

$P(C_2|\overline{C}_1)$ $\frac{13}{51}$ $P(\overline{C}_1 \cap C_2) = \frac{13}{68}$ 1 club

$P(\overline{C}_1)$

$\frac{39}{52}$

$P(\overline{C}_2|\overline{C}_1)$ $\frac{38}{51}$ $P(\overline{C}_1 \cap \overline{C}_2) = \frac{38}{68}$ 0 clubs

1

Without replacement:
Let C_1 be the event 'a club is drawn first'.
Let C_2 be the event 'a club is drawn second'.
From the tree diagram

(a) $P(C_2 | C_1) = 12/51 = 4/17$.

(b) $P(C_2) = P(C_1 \cap C_2) + P(\overline{C}_1 \cap C_2)$

$$= \left(\frac{13}{52}\right)\left(\frac{12}{51}\right) + \left(\frac{39}{52}\right)\left(\frac{13}{51}\right) = \frac{17}{68} = \frac{1}{4}.$$

Note: This would give the same answer for C_3, C_4, etc. The probability that any chosen card is a club is $\frac{1}{4}$.

(c) $P(0 \text{ clubs}) = \left(\frac{3}{4}\right)\left(\frac{38}{51}\right) = \frac{38}{68}$.

$P(1 \text{ club}) = \left(\frac{1}{4}\right)\left(\frac{13}{17}\right) + \left(\frac{3}{4}\right)\left(\frac{13}{51}\right) = \frac{26}{68}$,

$P(2 \text{ clubs}) = \left(\frac{1}{4}\right)\left(\frac{4}{17}\right) = \frac{4}{68}$.

Expected number of clubs $= \left(\frac{38}{68}\right)(0) + \left(\frac{26}{68}\right)(1) + \left(\frac{4}{68}\right)(2) = \frac{1}{2}$.

With replacement:

(a) Prob. $(C_2 | C_1 = \frac{1}{4}$

(b) Prob. $(C_2) = \frac{1}{4}$.

(c) $P(0 \text{ clubs}) = \left(\frac{3}{4}\right)\left(\frac{3}{4}\right) = \frac{9}{16}$.

$P(1 \text{ club}) = \left(\frac{13}{52}\right)\left(\frac{39}{52}\right)(2) = \left(\frac{6}{16}\right)$.

$$P(2 \text{ clubs}) = \left(\frac{1}{4}\right)\left(\frac{1}{4}\right) = \frac{1}{16}.$$

$$\text{Expected number of clubs} = \left(\frac{9}{16}\right)(0) + \left(\frac{6}{16}\right)(1) + \left(\frac{1}{16}\right)(2) = \frac{1}{2}.$$

12.5 A bag contains 6 white beads, 5 red beads and 4 blue beads. Three are selected at random without replacement. Find the probability that
(a) the third bead is white,
(b) the third bead is the first white,
(c) the beads were selected white, red and blue in that order,
(d) the three beads are different colours,
(e) the third bead is red, given that the first was red.
  Explain the difference between the expected number of white beads and the most likely number of white beads and find
(f) the expected number of white beads and
(g) the most likely number of white beads.

R.T.C. LIBRARY LETTERKENNY

- Total number of beads = 15.

(a) The probability that any bead is white = $\frac{6}{15}$.

(b) The probability that the first white bead is the third bead

$$= P(\text{1st not white}) \times P(\text{2nd not white}) \times P(\text{3rd is white})$$

$$= \left(\frac{9}{15}\right)\left(\frac{8}{14}\right)\left(\frac{6}{13}\right) = \frac{72}{455}.$$

(c) $P(\text{1st is white}) \times P(\text{2nd is red}) \times P(\text{3rd is blue}) = \left(\frac{6}{15}\right)\left(\frac{5}{14}\right)\left(\frac{4}{13}\right) = \frac{4}{91}.$

(d) White, red and blue can be arranged in $3! = 6$ different ways so

$$P(\text{3 different colours}) = (6)\left(\frac{4}{91}\right) = \frac{24}{91}.$$

(e) If the first bead is red then 6W, 4R and 4B beads are left.

$$P(\text{any bead is red}) = \frac{4}{14} = \frac{2}{7}. \qquad \text{Therefore } P(R_3 \mid R_1) = \frac{2}{7}.$$

The expected number of white beads is the statistical mean.
[In general $E(X) = \Sigma \{x\, P(X = x)\}$.]
  This is not usually an integer, i.e. not a member of the possibility space.
  The most likely number of white beads is that with the highest probability.

(f) Expected number of white beads $= 3\left(\frac{6}{15}\right) = \frac{6}{5}.$

(g) $P(\text{1 white bead}) = P(W, \overline{W}, \overline{W}) + P(\overline{W}, W, \overline{W}) + P(\overline{W}, \overline{W}, W)$

$$= \left(\frac{6}{15}\right)\left(\frac{9}{14}\right)\left(\frac{8}{13}\right)(3).$$

$P(\text{2 white beads}) = P(W, W, \overline{W}) \times 3$

$$= \left(\frac{6}{15}\right)\left(\frac{5}{14}\right)\left(\frac{9}{13}\right)(3).$$

Since $P(\text{1 W bead}) > P(\text{2 W beads})$, the most likely number of white beads is 1.

12.6 (a) Two boys play a game taking it in turns to roll a six-sided die. The first boy to obtain a six wins. What is the probability that the boy who rolls first wins on

  (i) his first roll,

 (ii) his second roll,

(iii) his third roll?

    What is the probability that the boy who rolls first wins?

    The rules of the game state that the boy who lost the first game should start the second game.

    What is the probability that the boy who rolled first in the first game wins both the first and the second game?

(b) A bag contains 5 black and 3 white balls and a second bag contains 3 black and 5 white. A ball is taken from the first bag and placed in the second, then a ball is taken from the second and placed in the first. A ball is now taken from the first bag. What is the probability that it is white?

● (a) (i) Prob. boy A wins on 1st roll = $\left(\dfrac{1}{6}\right)$.

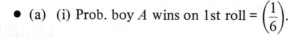

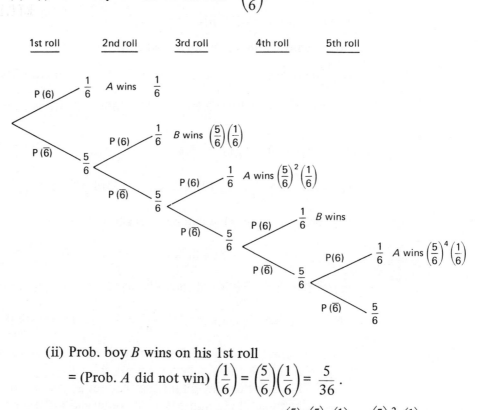

(ii) Prob. boy B wins on his 1st roll

$$= (\text{Prob. } A \text{ did not win}) \left(\frac{1}{6}\right) = \left(\frac{5}{6}\right)\left(\frac{1}{6}\right) = \frac{5}{36}.$$

Prob. boy A wins on his 2nd roll $= \left(\dfrac{5}{6}\right)\left(\dfrac{5}{6}\right)\left(\dfrac{1}{6}\right) = \left(\dfrac{5}{6}\right)^2\left(\dfrac{1}{6}\right)$.

(iii) Prob. boy A wins on his 3rd roll $= \left(\dfrac{5}{6}\right)^4 \left(\dfrac{1}{6}\right)$.

Prob. that boy A wins is $\dfrac{1}{6} + \dfrac{1}{6}\left(\dfrac{5}{6}\right)^2 + \left(\dfrac{1}{6}\right)\left(\dfrac{5}{6}\right)^4 + \ldots$.

This is a geometric progression with $a = \dfrac{1}{6}$, $r = \dfrac{25}{36}$.

Prob. that A wins $= \dfrac{a}{1-r} = \dfrac{1/6}{11/36} = \dfrac{6}{11}$.

If boy A wins the first game, boy B starts the second game and has

178

a probability of $\frac{6}{11}$ of winning the second game.

Thus boy A has a prob. of $\left(1 - \frac{6}{11}\right) = \frac{5}{11}$ of winning the second game.

Probability of boy A winning both games $= \left(\frac{6}{11}\right)\left(\frac{5}{11}\right) = \frac{30}{121}$.

(b) Let W_1 be the event 'ball drawn from first bag was white' etc.

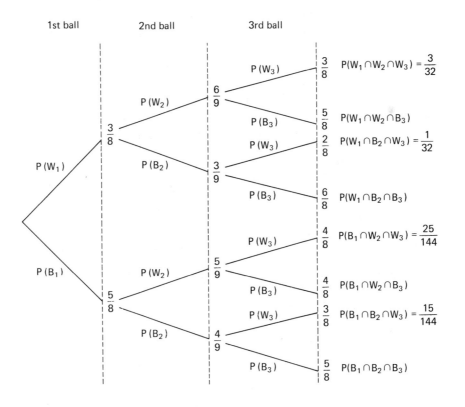

Probability that the final ball is white is
$$P(W_1 \cap W_2 \cap W_3) + P(W_1 \cap B_2 \cap W_3) + P(B_1 \cap W_2 \cap W_3) + P(B_1 \cap B_2 \cap W_3)$$
$$= \left(\frac{3}{8}\right)\left(\frac{6}{9}\right)\left(\frac{3}{8}\right) + \left(\frac{3}{8}\right)\left(\frac{3}{9}\right)\left(\frac{2}{8}\right) + \left(\frac{5}{8}\right)\left(\frac{5}{9}\right)\left(\frac{4}{8}\right) + \left(\frac{5}{8}\right)\left(\frac{4}{9}\right)\left(\frac{3}{8}\right)$$
$$= \frac{29}{72}.$$

12.7 (a) Two cards are drawn without replacement from ten cards which are numbered from 1 to 10. Find the probability that
 (i)   the numbers on both cards are even,
 (ii) the number on one card is odd and the number on the other card is even,
 (iii) the sum of the numbers on the two cards exceeds 4.
(b) Events A and C are independent. Probabilities relating to events A, B and C are as follows:

$$P(A) = \frac{1}{5}, \qquad P(B) = \frac{1}{6}, \qquad P(A \cap C) = \frac{1}{20}, \qquad P(B \cup C) = \frac{3}{8}.$$

Evaluate $P(C)$ and show that events B and C are independent.

● (a) (i) Probability of 2 even cards $= \left(\frac{1}{2}\right)\left(\frac{4}{9}\right) = \frac{2}{9}$.

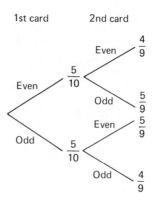

1st card   2nd card

(ii) Probability if 1 odd and 1 even $= \dfrac{5}{18} + \dfrac{5}{18} = \dfrac{5}{9}$.

(iii) Probability that the sum exceeds 4

$$= 1 - \{P(1, 2) + P(2, 1) + P(1, 3) + P(3, 1)\} = 1 - \left(\dfrac{1}{10}\right)\left(\dfrac{1}{9}\right)(4) = \dfrac{43}{45}.$$

(b) A and C are independent so $P(A \mid C) = P(A)$

Venn diagram

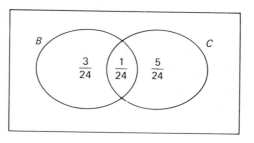

or $P(A) \cdot P(C) = P(A \cap C)$.

$P(A) = \dfrac{1}{5}$, $P(A \cap C) = \dfrac{1}{20}$,   thus $P(C) = \dfrac{1}{4}$.

$P(B) + P(C) = \dfrac{1}{6} + \dfrac{1}{4} = \dfrac{10}{24}$.

$P(B \cup C) = \dfrac{9}{24}$, therefore $P(B \cap C) = \dfrac{1}{24}$.

$P(B \mid C) = \dfrac{P(B \cap C)}{P(C)} = \dfrac{1/24}{1/4} = \dfrac{1}{6}$.

Therefore $P(B \mid C) = P(B)$. Hence B and C are independent.

12.3 Exercises

12.1 (multiple choice) There are six blue socks and four red socks in a drawer. When I go for a pair the light is fused, so the room is in darkness. What is the probability that if I choose two socks I will not end up with a pair?

A, $\dfrac{4}{15}$;   B, $\dfrac{12}{25}$;   C, $\dfrac{8}{15}$;   D, $\dfrac{29}{50}$;   E, none of these.

12.2   I need to replace the electric element in my kettle. In the nearest town there are 4 shops each of which have a probability of 0.75 of having the required element in stock. I try the shops in turn until I find the element (if one is available).

   Find   (a) the probability that I visit four shops,
              (b) the expected number of shops visited.

12.3   Nine beads, 5 red and 4 white are taken at random and threaded on a straight rod.
(a) What is the probability that no bead is next to one of the same colour?
(b) What is the probability that the last four beads include 2 red and 2 white beads?

12.4   Three apple trees and three pear trees are planted in a row in a random order.

   Find the probability that
(a) the three apple trees are together,
(b) the apple and pear trees are planted alternately.
What would be the corresponding probabilities if the trees are planted in a circle?

12.5   Explain what is meant statistically by the statement that two events, A and B, are independent.

   Let A be the event that the result of tossing a coin several times contains both heads and tails.

   Let B be the event that there is at most one head. Assuming that the probabilities of obtaining a head or tail are equal, determine whether the events A and B are independent if the coin is tossed
(a) three times,
(b) four times.

12.6   For events A and B, express $P(A \cup B)$ in terms of $P(A)$ and $P(B)$ only, when
(a) A and B are mutually exclusive,
(b) A and B are independent.
A bag contains just 10 balls, of which 5 are red and 5 are black. One ball is drawn at random from the bag and replaced; a second ball is drawn at random and replaced and then a third ball is drawn at random. By means of a tree diagram, or otherwise, show that the probability of drawing 2 black balls and one red ball is 3/8.

   Event A is that the 3 balls drawn include at least 1 red ball and at least 1 black ball. Event B is that the 3 balls drawn include at least 2 black balls. Find $P(A)$ and $P(B)$ and show that the events A and B are independent.

   Given that event C is that the first 2 balls drawn are of the same colour, ascertain whether events B and C are mutually exclusive.   (L)

12.7   (a) A and B play a game as follows: an ordinary die is rolled and if a six is obtained then A wins and if a one is obtained then B wins. If neither a six nor a one is obtained then the die is rolled again until a decision can be made. What is the probability that A wins on
(i) the first roll,   (ii) the second roll,   (iii) the rth roll?
What is the probability that A wins?
(b) A bag contains 4 red and 3 yellow balls and another bag contains 3 red and 4 yellow. A ball is taken from the first bag and placed in the second, the second bag shaken and a ball taken from it and placed in the first bag.

If a ball is now taken from the first bag what is the probability that it is red? (You are advised to draw a tree diagram.) (SUJB)

12.8 A boat hiring company at the local boating lake has two types of craft; 20 Bluebirds and 35 Herons. The customer has to take the next boat available when hiring. The boats are all distinguishable by their numbers.

A regular customer, John, carefully notes the numbers of the boats which he uses, and finds that he has used 15 different Bluebirds and 20 different Herons. Each boat is equally likely to be the next in line. John hires two boats at the same time (one is for a friend).

If event X is: John has not hired either boat before; and event Y is: both boats are Herons,

determine   (a) $P(X)$;   (b) $P(Y)$;   (c) $P(X \mid Y)$;   (d) whether the events are mutually exclusive.

12.4 Brief Solutions to Exercises

12.1   Probability of odd socks $= P(B_1 \cap R_2) + P(R_1 \cap B_2)$

$$= \left(\frac{6}{10}\right)\left(\frac{4}{9}\right) + \left(\frac{4}{10}\right)\left(\frac{6}{9}\right) = \left(\frac{8}{15}\right).$$

Answer C

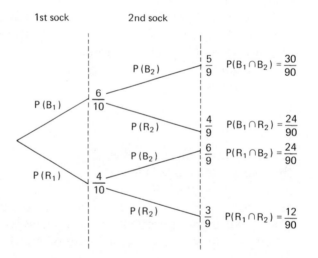

12.2   (a)   I visit 4 shops if first 3 do not have an element.

$$P(\text{visit 4 shops}) = \left(\frac{1}{4}\right)^3 = \frac{1}{64}.$$

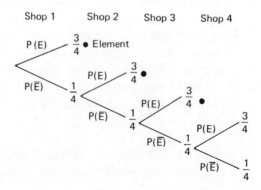

(b) P (only 1 visited) = $\dfrac{3}{4}$.

P (only 2 visited) = $\left(\dfrac{1}{4}\right)\left(\dfrac{3}{4}\right) = \dfrac{3}{16}$.

P (only 3 visited) = $\left(\dfrac{1}{4}\right)^2\left(\dfrac{3}{4}\right) = \dfrac{3}{64}$.

P (4 visited) = $\left(\dfrac{1}{4}\right)^3 = \dfrac{1}{64}$.

Expected number = $1\left(\dfrac{3}{4}\right) + 2\left(\dfrac{3}{16}\right) + 3\left(\dfrac{3}{64}\right) + 4\left(\dfrac{1}{64}\right)$

$= \dfrac{85}{64}$.

12.3 5 red and 4 white can be arranged in $\dfrac{9!}{4!\,5!} = 126$ ways.

Only one is RWRWRWRWR, therefore

(a) P (no bead is next to one of the same colour) = $\dfrac{1}{126}$.

(b) Probability that any group of 4 selected are RRWW = $\left(\dfrac{5}{9}\right)\left(\dfrac{4}{8}\right)\left(\dfrac{4}{7}\right)\left(\dfrac{3}{6}\right) = \dfrac{5}{63}$.

Number of arrangements of 2 red and 2 white

$= \dfrac{4!}{2!\,2!} = 6, \quad \Rightarrow \quad$ probability is $\dfrac{10}{21}$.

12.4 (a) P (AAAPPP) = $\left(\dfrac{3}{6}\right)\left(\dfrac{2}{5}\right)\left(\dfrac{1}{4}\right)(1) = \dfrac{1}{20}$.

Arrangements of (AAA), P, P, P = 4.

Therefore prob. of 3 apples together = $\dfrac{1}{20}\,4 = \dfrac{1}{5}$.

(b) P (APAPAP) + P (PAPAPA) = $\left(\dfrac{3}{6}\right)\left(\dfrac{3}{5}\right)\left(\dfrac{2}{4}\right)\left(\dfrac{2}{3}\right)\left(\dfrac{1}{2}\right)\left(\dfrac{1}{1}\right)(2) = \dfrac{1}{10}$.

In a circle fix one tree (A) and arrange the others. 2A and 3P can be arranged in $\dfrac{5!}{2!\,3!} = 10$ ways.

A, A, (PPP) can be arranged in 3 ways.

(a) P (apples together) = $\dfrac{3}{10}$.

P (trees planted alternately) = $\dfrac{1}{10}$.

12.5 See Fact Sheet, section 12.1.

(a) P $(A) = \dfrac{3}{4}$, P $(B) = \dfrac{1}{2}$.

P $(A \mid B) = \dfrac{P(1H, 2T)}{P(B)} = \dfrac{3}{4} = $ **P** (A) thus independent.

(b) $P(A) = \dfrac{7}{8}$, $P(B) = \dfrac{5}{16}$.

$$P(A \mid B) = \frac{P(1H, 3T)}{P(B)} = \frac{4}{5} \neq P(A) \text{ thus not independent.}$$

12.6 (a) and (b) see Fact Sheet, section 12.1 (e), (d) and (b)(iii).

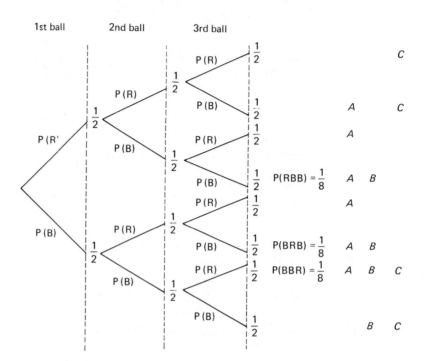

$$P(B, B, R) \times \text{no. of arrangements} = \left(\frac{1}{2}\right)^3 3 = \frac{3}{8}.$$

$$P(B, R, R) \times \text{no. of arrangements} = \left(\frac{1}{2}\right)^3 3 = \frac{3}{8}.$$

$$P(A) = \frac{3}{8} + \frac{3}{8} = \frac{3}{4}.$$

$$P(B) = P(2B \text{ and } 1R) + P(3B) = \frac{3}{8} + \frac{1}{8} = \frac{1}{2}.$$

$$P(A \mid B) = \frac{P(A \cap B)}{P(B)} = \frac{3/8}{4/8} = \frac{3}{4} = P(A) \quad \Rightarrow \quad A \text{ and } B \text{ are independent.}$$

$$P(C) = \frac{1}{2}, \quad P(C \cap B) = \frac{1}{4},$$

$$P(C \cap B) \neq 0 \quad \Rightarrow \quad B \text{ and } C \text{ are not exclusive.}$$

12.7 (a) (i) $\dfrac{1}{6}$, (ii) $\left(\dfrac{1}{6}\right)\left(\dfrac{4}{6}\right)$, (iii) $\left(\dfrac{1}{6}\right)\left(\dfrac{4}{6}\right)^{r-1}$.

Probability that A wins $= \dfrac{1}{6}\left(1 + \left(\dfrac{4}{6}\right) + \left(\dfrac{4}{6}\right)^2 + \ldots\right) = \dfrac{1}{2}.$

(b)

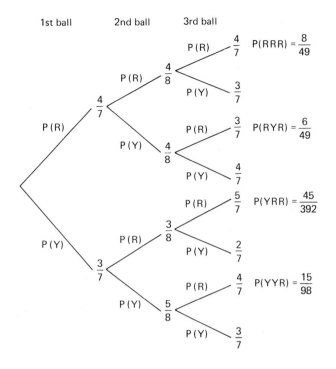

1st ball   2nd ball   3rd ball

$P(R)$ $\frac{4}{7}$ $P(RRR) = \frac{8}{49}$

$P(R)$ $\frac{4}{8}$

$P(Y)$ $\frac{3}{7}$

$\frac{4}{7}$

$P(R)$

$P(R)$ $\frac{3}{7}$ $P(RYR) = \frac{6}{49}$

$P(Y)$ $\frac{4}{8}$

$P(Y)$ $\frac{4}{7}$

$P(R)$ $\frac{5}{7}$ $P(YRR) = \frac{45}{392}$

$\frac{3}{8}$

$P(R)$

$P(Y)$ $\frac{2}{7}$

$P(Y)$

$\frac{3}{7}$

$P(R)$ $\frac{4}{7}$ $P(YYR) = \frac{15}{98}$

$P(Y)$ $\frac{5}{8}$

$P(Y)$ $\frac{3}{7}$

Probability that ball is red is

$$\left(\frac{4}{7}\right)\left(\frac{4}{8}\right)\left(\frac{4}{7}\right) + \left(\frac{4}{7}\right)\left(\frac{4}{8}\right)\left(\frac{3}{7}\right) + \left(\frac{3}{7}\right)\left(\frac{3}{8}\right)\left(\frac{5}{7}\right) + \left(\frac{3}{7}\right)\left(\frac{5}{8}\right)\left(\frac{4}{7}\right) = \frac{31}{56} .$$

12.8   (a) P (has not hired 1st boat) $= \frac{20}{55}$,

P (has not hired 2nd boat) $= \frac{19}{54}$,

$\Rightarrow$ $P(X) = \left(\frac{20}{55}\right)\left(\frac{19}{54}\right) = \frac{38}{297}$.

(b) P (1st boat is a Heron) $= \frac{35}{55}$,

P (2nd boat is a Heron) $= \frac{34}{54}$,

$\Rightarrow$ $P(Y) = \left(\frac{35}{55}\right)\left(\frac{34}{54}\right) = \frac{119}{297}$.

(c) $P(X|Y) = \dfrac{P(X \cap Y)}{P(Y)}$, where $P(X \cap Y) = \left(\frac{15}{55}\right)\left(\frac{14}{54}\right) = \frac{7}{99}$.

Therefore $P(X|Y) = \dfrac{7/99}{119/297} = \dfrac{3}{17}$.

(d) $P(X \cap Y) \neq 0$ $\Rightarrow$ events are not mutually exclusive.

Index